ALLE ZEIT WACH 1842

H. Felix G. Haemmerli P. Sträuli

Dynamic Morphology of Leukemia Cells

A Comparative Study by
Scanning Electron Microscopy
and Microcinematography

With 111 Figures

Springer-Verlag
Berlin Heidelberg New York 1978

Dr. Heidi Felix
Dr. Gisela Haemmerli
Prof. Dr. Peter Sträuli

Division of Cancer Research
Institute of Pathology, University of Zürich
Birchstraße 95, CH-8050 Zürich (Switzerland)

This study was performed within the Leukemia Working Group of the University of Zürich sponsored by the Zürich Cancer League.

ISBN-13:978-3-642-66796-1 e-ISBN-13: 978-3-642-66794-7
DOI: 10.1007/978-3-642-66794-7

Library of Congress Catalog Card Number: Felix, Heidi, 1941–. Dynamic morphology of leukemia cells. Bibliography: p. Includes index. 1. Leukemia. 2. Cancer cells. 3. Cinematography, Medical. 4. Scanning electron microscope. 5. Rats-Diseases. I. Haemmerli, G., 1923– joint author. II. Sträuli, P., 1918– joint author. III. Title. [DNLM: 1. Leukemia, Experimental – Physiopathology. WH250 F316d]
RC643.F44 616.1'55 77-14964

Softcover reprint of the hardcover 1st edition 1978

Reproduction of the figures: Brend'amour, Simhart & Co., Munich
2121/3140-543210

To our friend Marcel Bessis

Contents

Introduction

Dynamic Morphology is the attempt to correlate surface architecture and shape of fixed cells, as visualized by scanning electron microscopy (SEM), with the behavior of living cells, recorded by microcinematography (MCM). If SEM and MCM are used concurrently for the analysis of cell populations, a dynamic interpretation of SEM photographs is only valid if the experimental conditions are identical for the two techniques. This is achieved by allowing the cells to settle on a glass surface where they remain long enough to perform their various activities under conditions identical for both techniques (for technical details see Methodology).

The analysis of a population necessitates the study of a large number of cells.
This prerequisite is met by operating the scanning electron microscope at low levels of magnification, and by using culture chambers for cinematography. It can be argued that the examination of attached cells excludes a complete SEM survey of a population, as cells not adhering from the outset or becoming detached during the different preparatory steps are lost. For this, cinematography proved to be a reliable control: All cell types recognized in time-lapse films were also seen in scanning electron (SE) micrographs. Another, and more general, objection to a dynamic interpretation concerns the artificiality of cellular behavior on glass. This is true, but does not invalidate comparative studies making use of this substrate. Moreover, we have also recorded the behavior of rat leukemia cells on a natural substrate, isologous mesentery maintained under organ-culture conditions. Despite the more complex substrate, the motility of the leukemia cells is identical to that displayed on glass. There is also a striking similarity of shape and surface of these leukemia cells when viewed by SEM on the glass surface and in bone-marrow preparations of leukemic animals.

Bessis and de Boisfleury used the combined approach of SEM and MCM as early as 1971 for the description of white blood cell movement and extended this study to leukemia cells in 1976. Haemmerli and Felix (1976) reported on the close correlation between the dynamic state of human acute leukemia cells and their surface morphology. In this book, surface architecture, shape, and motile behavior of cells from two rat leukemias and from 19 acute and chronic human leukemias of myeloid, lymphoid, and unclassifiable nature is presented. Hematologic data from the human leukemias and technical details are given in the appendix. Some transmission electron (TE) micrographs are included to provide supplementary information or to present the thin-section aspect of characteristic configurations.

Dynamic morphology requires an appropriate nomenclature. We have selected a terminology that only takes into account the basic features of cell surface architecture and cell shape.
With regard to ***Surface Architecture*** the variability of cytoplasmic extensions appears enormous. We believe, however, that only three basic structures are involved in this diversity: microvilli, folds, and blebs.
Microvilli can be shorter or *stublike* and longer or *fingerlike*.
Folds comprise all planar extensions. Their extremes are the low *ridges* and the large, thin *veils*. Pleated folds, the momentary appearance of these extensions fixed during undulation, are called *ruffles*.
Blebs are round, short extensions mostly described in conjunction with cell division and agony.

Contact to the substrate is established by *attachment filaments*, thin projections of various lengths,and by *attachment plates* that connect the cell's with a flat, broad cytoplasmic base to the glass.

Size, number, and distribution of the different cytoplasmic structures are subject to changes in accordance with cellular activity.

As concerns the ***Shape*** of the cells, the terms spheric and polarized are used.
Most *spheric* cells show surface motility expressed as rapid protrusion and retraction of cytoplasmic extensions. It is performed

while the cells remain sessile. In our experience, SEM cannot distinguish between different cell types as long as they are spheric. This is mainly due to the variability in surface features of leukemia cells within the same population. These differences are the expression of the surface motility that subjects the various cytoplasmic extensions to continuous change. The aspect in SE micrographs corresponds to the momentary state at the time of fixation.

There is one exception, and that is the monocyte. When monocytes are spheric, they are covered with large, thin folds which are different from the surface extensions of the other white blood cells. In addition, monocytes, given the necessary time, have the tendency to spread, a special feature not seen for the other white-cell types in our preparations. This finding, however, is in contrast to reports where, for instance, the spreading capacity of granulocytes was noted (Michaelis et al., 1971; Bessis and de Boisfleury, 1971; Allen and Dexter, 1976).

A cell is *polarized* if it has two opposite, morphologically distinct parts. Polarization is indicative for either locomotion, the form of motility connected with translocation of the whole cell, or on-spot motility. With some exceptions, SEM can distinguish between the two types of motility by the analysis of contact zones.

If a polarized cell adheres to the substrate or to another cell by means of a footlike extension without further contact areas, it can be assumed that the activity is of the on-spot type. In contrast, locomotive cells have not only contact zones on the posterior part, but also on the body and the front part. The posterior part can have the form of a tail. This, however, is not a prerequisite of the polarized state. In the case of locomotive myelocytes, for instance, polarization is achieved by the extension of veils in the direction of movement. This can be seen particularly well in SE micrographs.

On what basis, then, can the analysis of a leukemia-cell population be carried out? The concurrent use of SEM and MCM allows the identification of the different cell types in SE micrographs, provided that the cells have been fixed while engaged in a mode of motility requiring polarization.

Blast cells have their characteristic manner of locomotion. They move in a very distinct, polarized configuration with a tail at the end and a roundish anterior part with various-sized extensions. Similar observations were reported by Boll and Nitzel (1972), Boll (1976), de Bruyn (1944, 1946), Norberg et al. (1973, 1974), and Tchernia et al. (1976), while Rich and co-workers (1939) and Senda et al. (1961) restricted the polarized mode of locomotion to lymphoblasts.
Dynamic morphology cannot distinguish between blast cells of different origin. There is no individually characteristic surface morphology, nor is there a difference in the mode of locomotion: all blast cells move in what we call „blast pattern of locomotion" (Haemmerli and Felix, 1977).

The typical appearance of the large locomotive *promyelocyte* in time-lapse films with its granules and elongated body can be recognized in SE micrographs without difficulty. SEM, in addition, furnishes information not visible in phase contrast, concerning, for instance, the tendency of promyelocytes to lift part of their body off the substrate.

Myelocytes are especially easy to recognize by SEM during their on-spot and locomotive activity. They have a round body and large veils which are arranged around the cell body in case of on-spot motility, and extended into the direction of movement on locomotive cells. Thus, the polarized aspect of myelocytes is completely different from that of blast cells and promyelocytes.

How do the possibilities of dynamic morphology compare with those reported in static SEM studies of leukemia cells? It is evident that the basis for comparison, due to differences in the technical approach, is small. Most reports describe cells that were fixed in suspension, a state corresponding to transport in the blood. It can be argued that transport offers suitable conditions for comparing different cell types, as all are in the same functional state. But this may not be completely true. Although transported cells are inactive as far as adhesion and locomotion are concerned, their surface motility cannot, a priori, be assumed to be identical. At any rate, it must be kept in mind that the surface architecture of "circulating" leukemia cells, as described by several authors (Bessis, 1973; Golomb et al., 1975a, b; Polliack et al., 1975a; Coleman et al., 1976; Deegan et al., 1976; Golomb and Reese, 1976; Polliack, 1976; Polliack et al., 1976), reflects only one facet out of a broad range of morphologic expressions.

In view of this variability we are sceptical about the possibility of distinguishing leukemia cells of T- and B-cell nature on the basis of their surface architecture (Polliack et al., 1975b; Polliack, 1976). We support the view that differences in surface morphology of T and B cells are not sufficiently distinct (Catovsky et al., 1975; Haemmerli et al., in press), and that parallel immunologic tests are needed for the identification of different types of lymphoid leukemias (Polliack and de Harven, 1975; Belpomme et al., 1976; Dantchev and Belpomme, 1977; Müller-Hermelink and Müller-Hermelink, 1977). We are also of the opinion that existing differences in surface morphology reflect transitory functional states rather than stable surface features of either T or B cells (Alexander et al., 1976; Cohnen et al., 1976).

The photographs in this book demonstrate the morphologic versatility of leukemia cells. This versatility, however, is not unlimited, and dynamic morphology can help us to seize the functional principles underlying the multitude of structural manifestations. This, at least, is true of cell shape. There are a few prototypes that distinct classes of leukemia cells assume with great tenacity whenever they are induced to abandon the spheric configuration. The functional triggers are locomotion and adhesion, and the inclusion of these cell activities in our evaluation adds a new dimension to our understanding of leukemia cells. For surface architecture, the situation is less clear-cut than for shape: The display of morphologic patterns at the cell surface is extremely varied, and although, in our opinion, it is traceable to a few basic structures, its functional background is far from being understood. We can only say that in a general way it reflects the cells' struggle for existence within their natural or artificial milieu. As long as the biology of this encounter is not explored, attempts at cell classification based on surface architecture will always be fraught with the danger of misinterpretation. It is—to quote Bessis (1976)—"hazardous to base claims for physiologic responses on static pictures arranged in sequence by the author rather than by nature."

I. Rat Leukemias

Material

1. BN-ML: Myeloid Leukemia of the BN Rat

This leukemia originated in an inbred BN/Rij rat following treatment with DMBA [van Bekkum et al., 1976; Colly et al. (personal communication); Hagenbeek, 1977; Hermens et al. (personal communication); Martens and Hagenbeek (personal communication)]. The promyelocytic nature of the cells is shown by the usual hematologic methods including various cytochemical reactions and by our TEM studies. In our laboratory, cells from leukemic animals were used for establishing an ascites tumor line growing exclusively in inbred BN rats. Intravenous and intraperitoneal injection of 10^7 tumor cells is followed by death of the animals, after approximately 3 weeks, from widespread leukemia.

Analysis of time-lapse films shows the BN leukemia cells to be essentially nonlocomotive, motility being restricted to a pronounced surface activity.

2. L 5222: Unclassifiable Leukemia of the BD IX Rat

L 5222 was induced by diethylnitrosourea in the inbred BD IX rat [Harriss and Hoelzer, 1971; Hoelzer et al., 1973, 1974; Harriss and Hoelzer, 1974; Ivankovic and Zeller, 1974; Rasche et al., 1974; Zeller et al., 1975; Fiebig et al., 1976]. The leukemia is considered unclassifiable, as neither cytochemistry nor TEM permit certain identification of the cells as belonging to either the myeloid or the lymphoid line.

Serial interperitoneal implantation of cells from leukemic animals yielded an ascites tumor line. Irrespective of the site of implantation, animals implanted with 10^7 L 5222 cells die between days 6 and 7 from a leukemia involving all organs.

In contrast to BN-ML cells, those of the L 5222 leukemia are highly locomotive [Haemmerli et al., 1976]. Furthermore, microcinematography revealed the ability of these leukemia cells to change their shape from round to completely flat, thus permitting them to crawl under other, spread cells [Haemmerli et al., 1977]. The importance of locomotion and shape adaptation during the process of organ infiltration was verified under tissue-culture conditions [Pourreau et al., 1977].

The comparative evaluation of the L 5222 and the BN-ML rat leukemia models and their relevance for human acute leukemia was the object of a 1976 workshop [Hagenbeek and van Bekkum, 1977].

The myeloid leukemia of the BN rat in its intraperitoneal form of growth consists of a population of uniformly sized tumor cells and a few macrophages from the host peritoneal cavity. Microcinematography shows that the tumor cells are essentially nonlocomotive. It is very rare that a tumor cell can be observed while migrating slowly for a short distance. The activity of the cells is restricted to pronounced surface motility that is expressed in rapid projection and retraction of small cytoplasmic extensions. In accordance with this activity, BN-ML cells are spheric.

Figure 1

Both SEM (a) and phase contrast from time-lapse films (b) show the spheric shape of BN-ML cells

a SEM × 3200
b Phase contrast × 3000

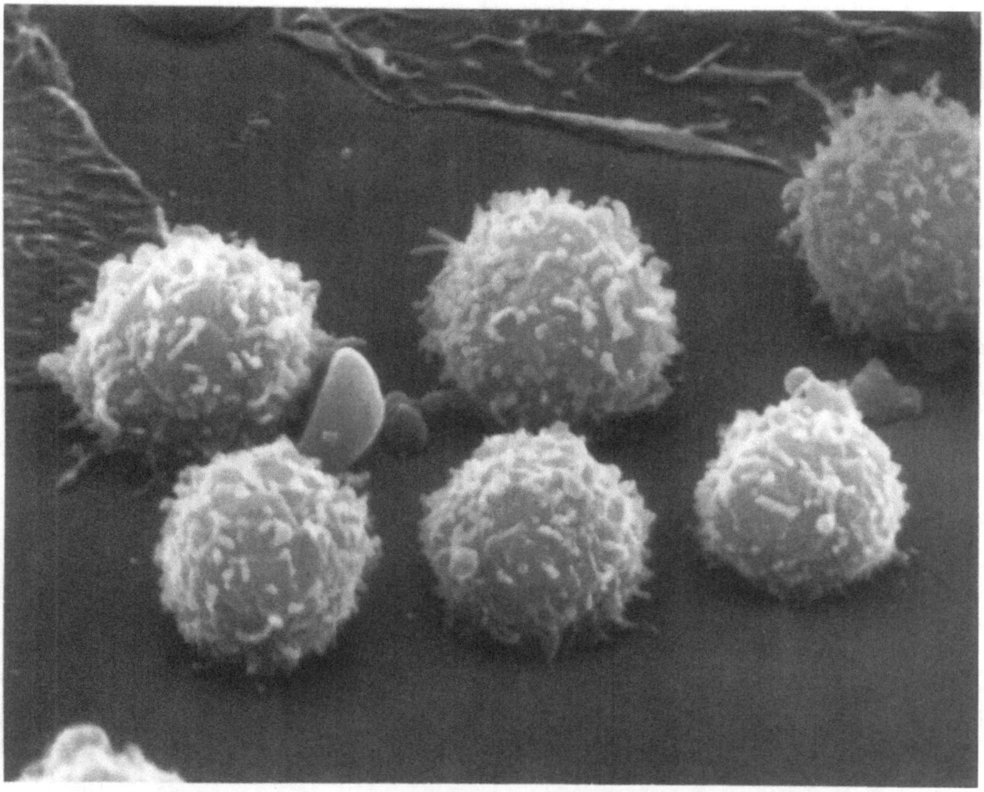

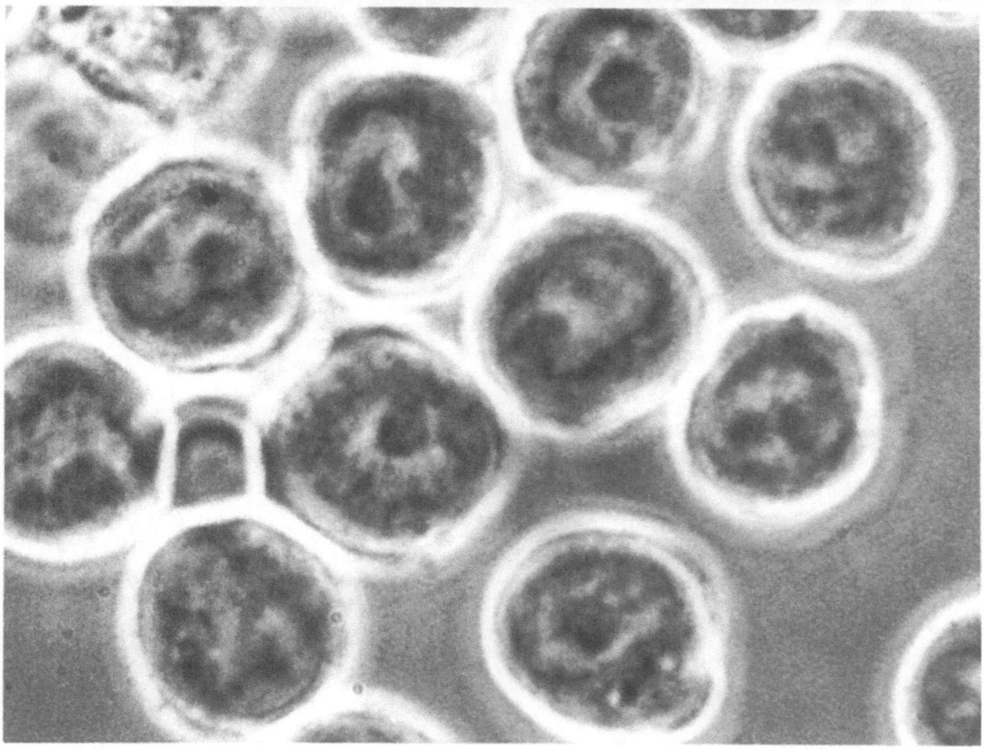

Figure 2

This SEM photograph (a) *reveals that the surface not only contains microvilli (mv) but also blebs (b) and folds (f). This corresponds to the uneven contours in the TE micrograph* (b). *Based on time-lapse film analyses these surface features can be interpreted as the result of surface motility. In the right-hand corner of the SEM picture, part of a half-spread macrophage (ma) is visible*

a SEM: ×6400
b TEM: ×5600

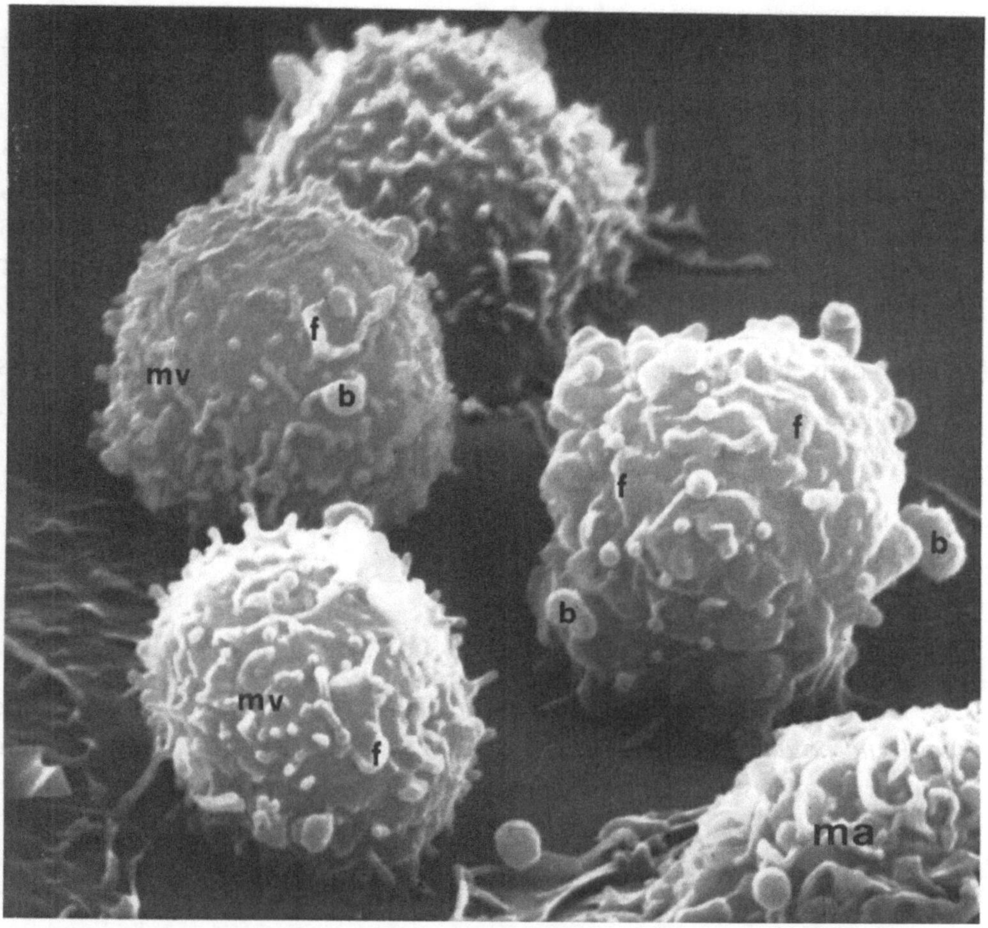

a

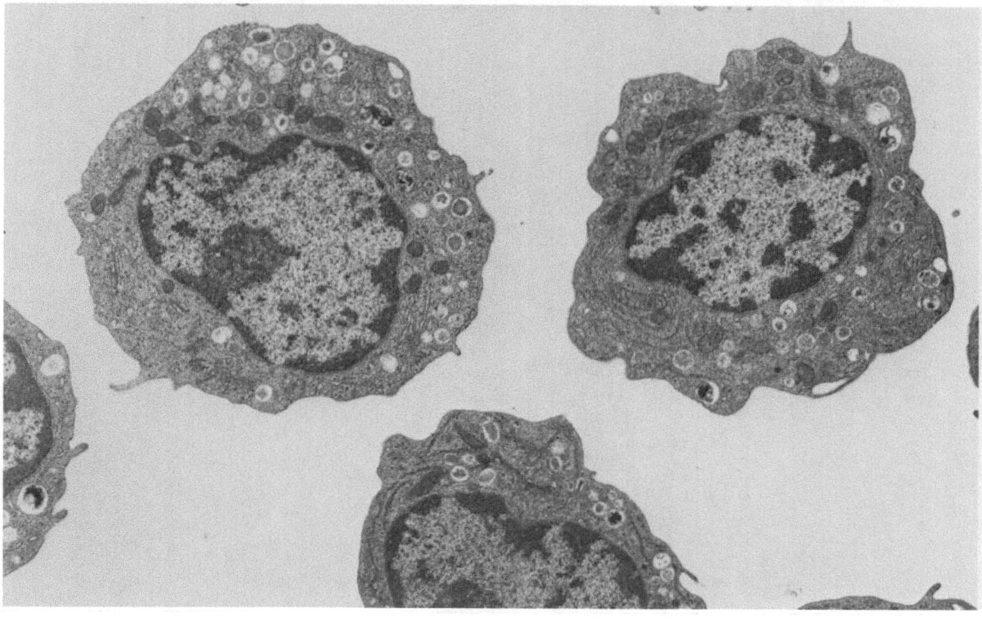

b

A survey of the population of the undifferentiated L 5222 leukemia of the BD IX rat shows that, in contrast to the BN-ML leukemia, many cells have assumed a polarized configuration. Time-lapse films have shown that this shape is the result of either on-spot motility or of locomotion.

Spheric cells are either at rest or show surface motility that is less pronounced than that of the BN leukemia cells. Rest, on-spot motility, and locomotion are interchangeable.

Figure 3

The Nomarski contrast photograph from a time-lapse film shows the round and polarized configuration of L 5222 cells. This still picture alone does not allow us to distinguish whether the polarized cells are engaged in on-spot motility, or in locomotion

Nomarski contrast: × 1600

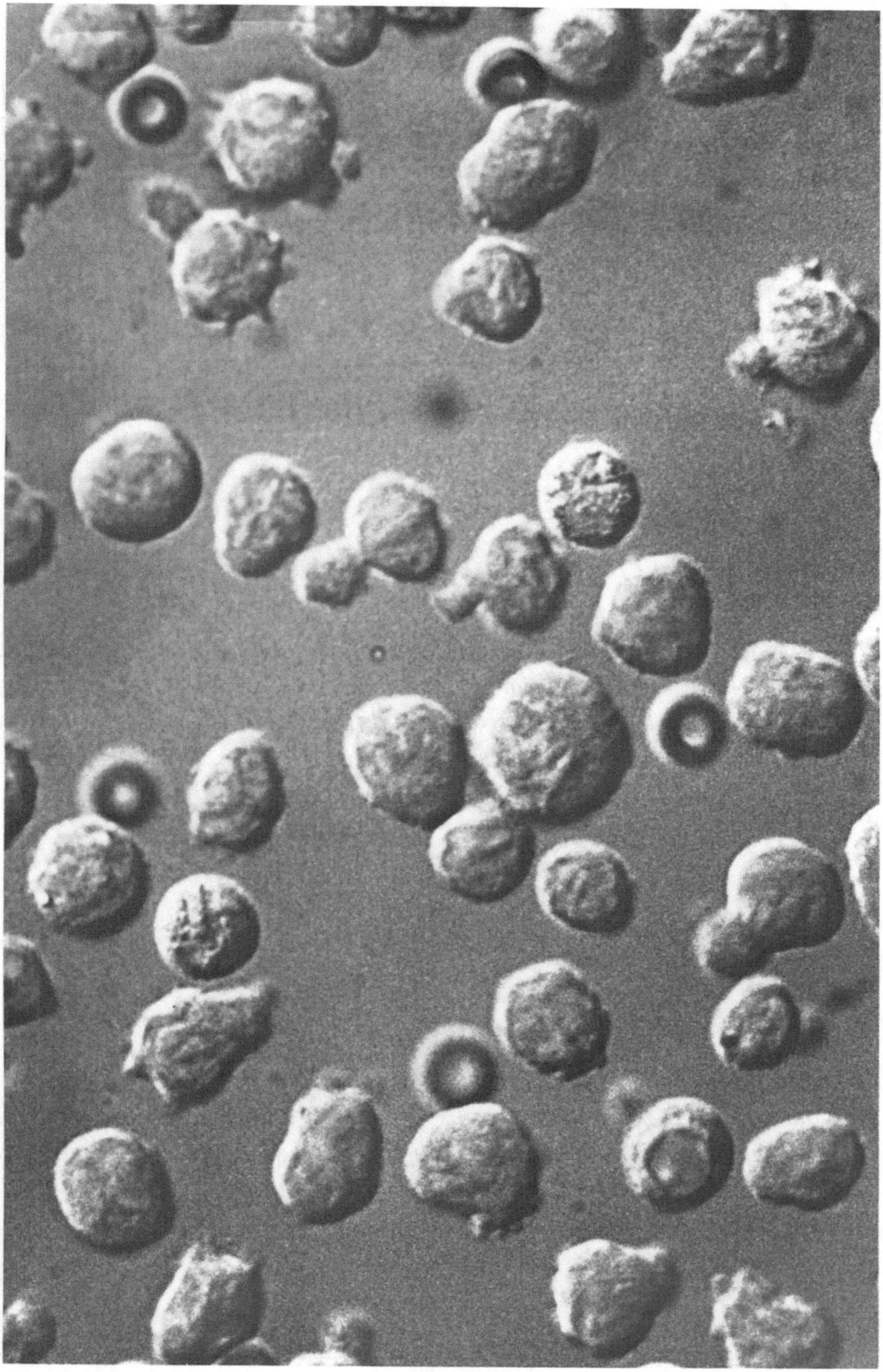

Figure 4

a *Two spheric and a polarized L 5222 cell are presented, together with a half-spread monocyte (mo). While the spheric cells are evenly covered with small microvilli, the polarized cell with its prominent tail and large veils at the front has a different distribution of the microvilli: they are now concentrated at the tail*

SEM: × *7300*

b *A higher magnification of a spherical L 5222 cell shows the uniform distribution of the short microvilli*

SEM: × *14,000*

Figure 5 *(see page 12)*

These L 5222 cells were fixed while displaying on-spot motility in the polarized shape. The posterior extension is used as a foot for making contact with the glass surface. Note the difference in surface morphology compared to that of spheric cells (Figure 4)

a SEM: × *13,900*
b SEM: × *9800*

Figure 6 *(see page 13)*

A spread macrophage from the rat peritoneal cavity serves as substrate for the two polarized leukemia cells. While the one in an upright position could be engaged in on-spot activity, the bent cell seems to have established additional contact with its front part. Therefore, the cell could be in the process of becoming locomotive

SEM: × *12,200*

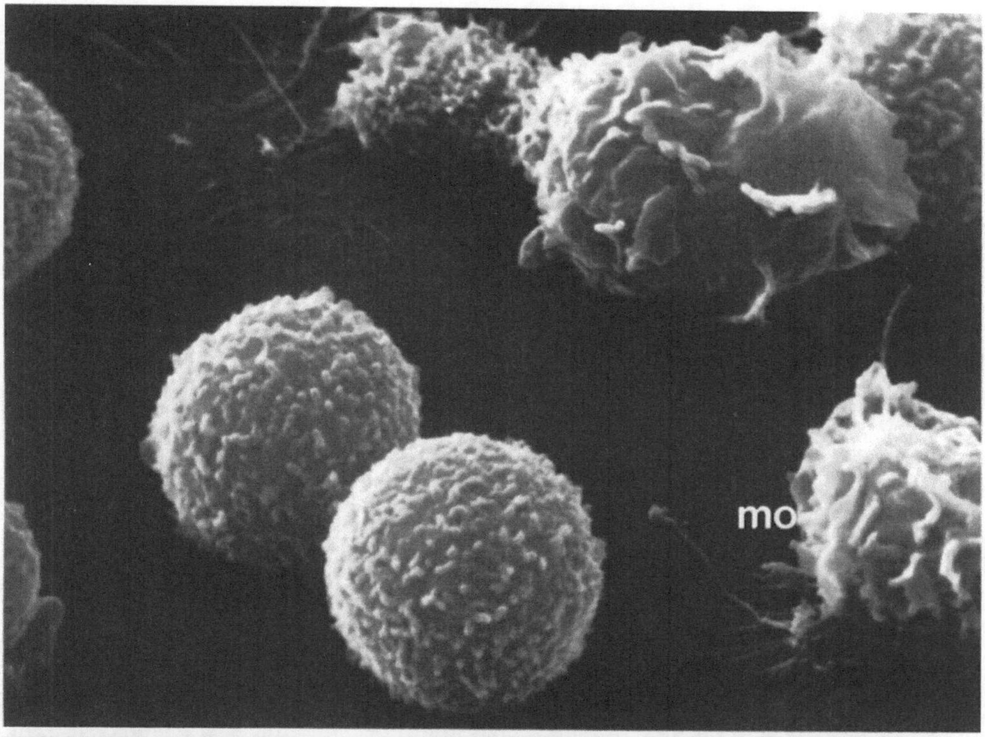

a

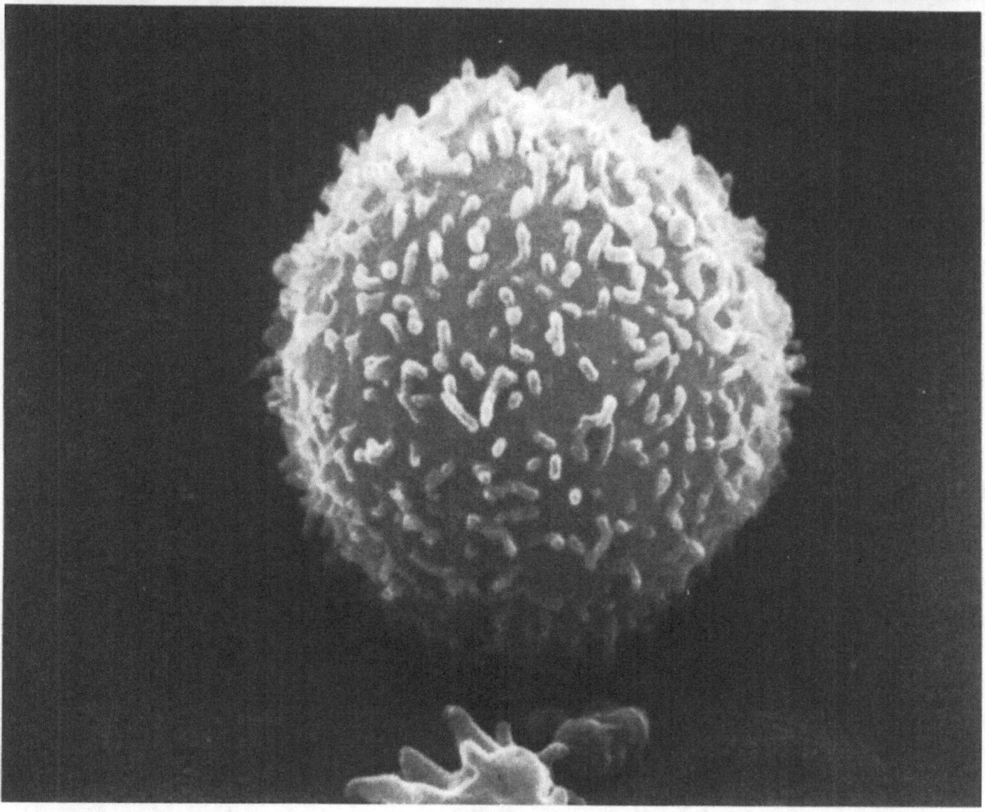

b

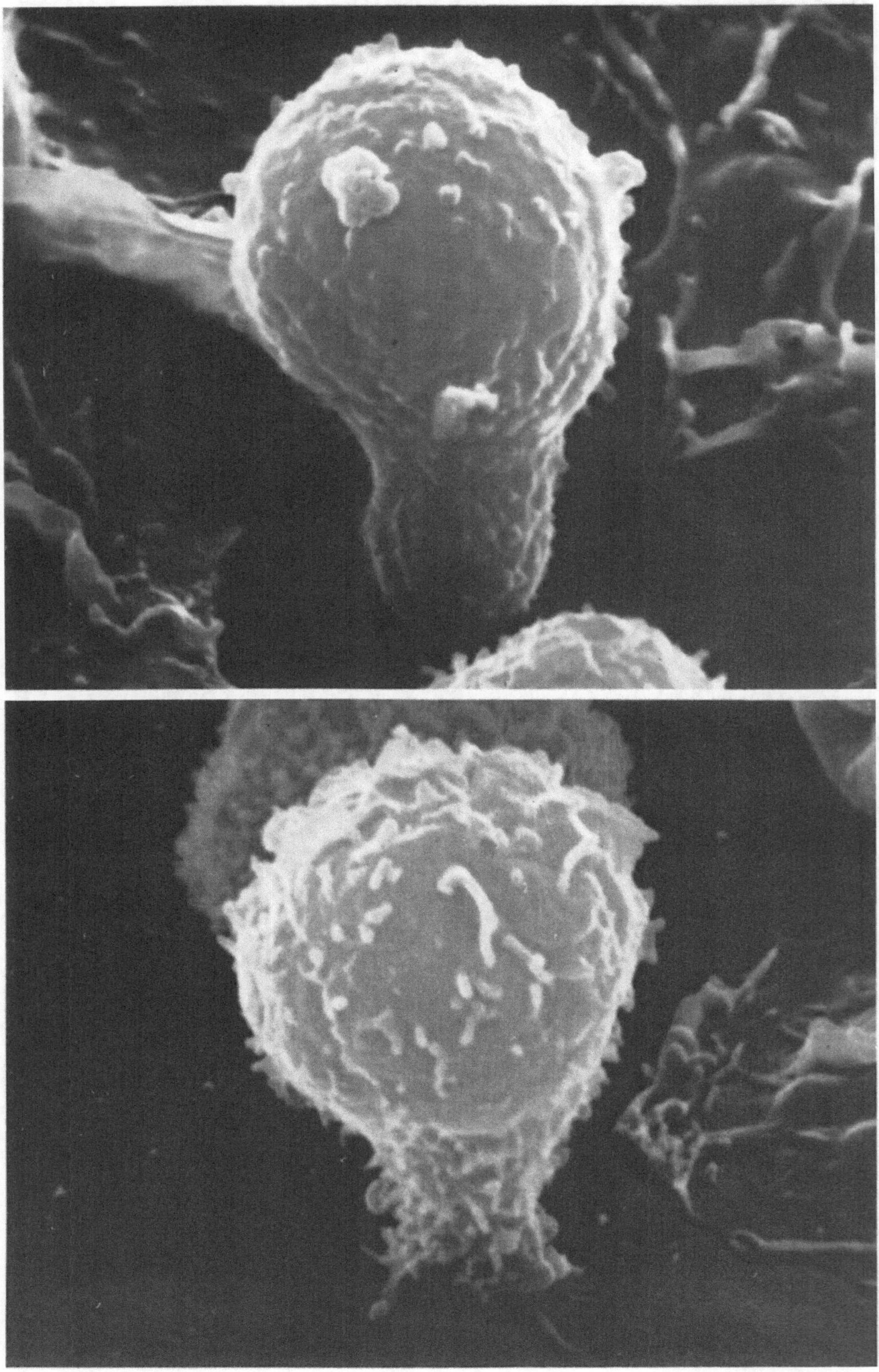

a

b

12

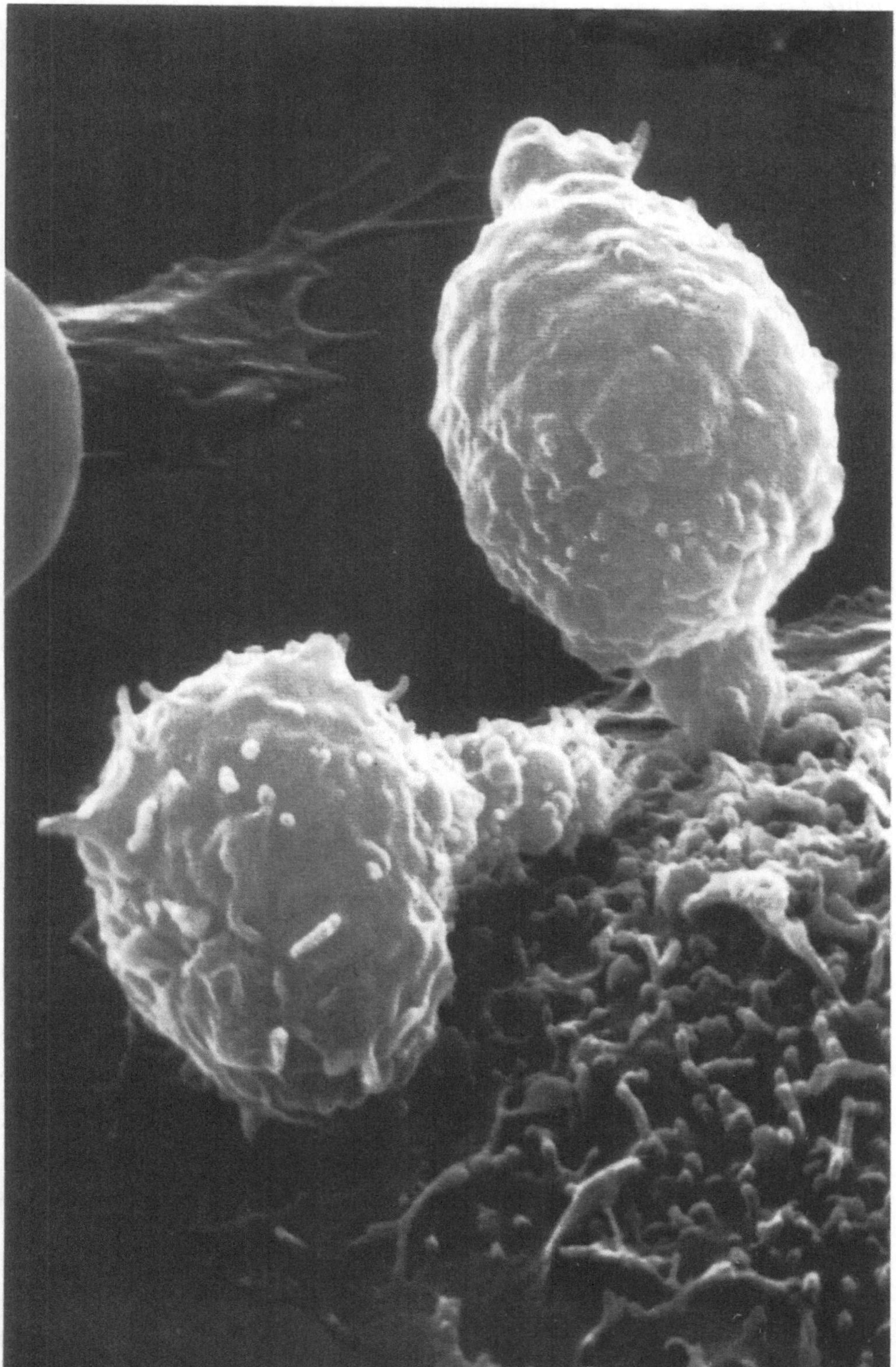

Figure 7

This series illustrates our concept of the transition from on-spot motility to locomotion in polarized L 5222 cells. From the initial posture (a), the cell gradually bends down (b, c), and then establishes contact between its body and the substrate (d). This is the configuration of a locomotive cell, in which the former top is now the leading part, and the foot the tail (e). Characteristic for locomotive L 5222 cells is their strict polarization, together with the well-developed veils at the leading part. Note that the cell has established contact to the substrate with its anterior part, its body and the tip of the tail. The curvature of the rest of the tail is a frequent finding. The direction of movement is indicated by an arrow

a SEM: × 8600
b SEM: × 8000
c SEM: × 7600
d SEM: × 6700
e SEM: × 9700

Figure 8 *(see page 16)*

Many polarized locomotive cells are present in this phase-contrast picture from a time-lapse film. The image of the L 5222 cell (arrow) performing on-spot motility is blurred as the cell has extended its top into the medium. There are also two myelocytes (my) and part of a spread macrophage (ma) from the peritoneal cavity of the host

Phase contrast: × 1600

Figure 9 *(see page 17)*

a This locomotive L 5222 has made contact with the glass substrate by means of a large anterior cytoplasmic extension, some smaller folds on the body, and a few microvilli on the tip of the tail (arrows). The middle portion of the tail is lifted off the substrate (see Figure 7e)

b The polarized configuration is also present in the TE micrograph. While the nucleus is located in the main body of the cell, most organelles of the cytoplasm are in the tail portion

a SEM: × 11,300
b TEM: × 11,800

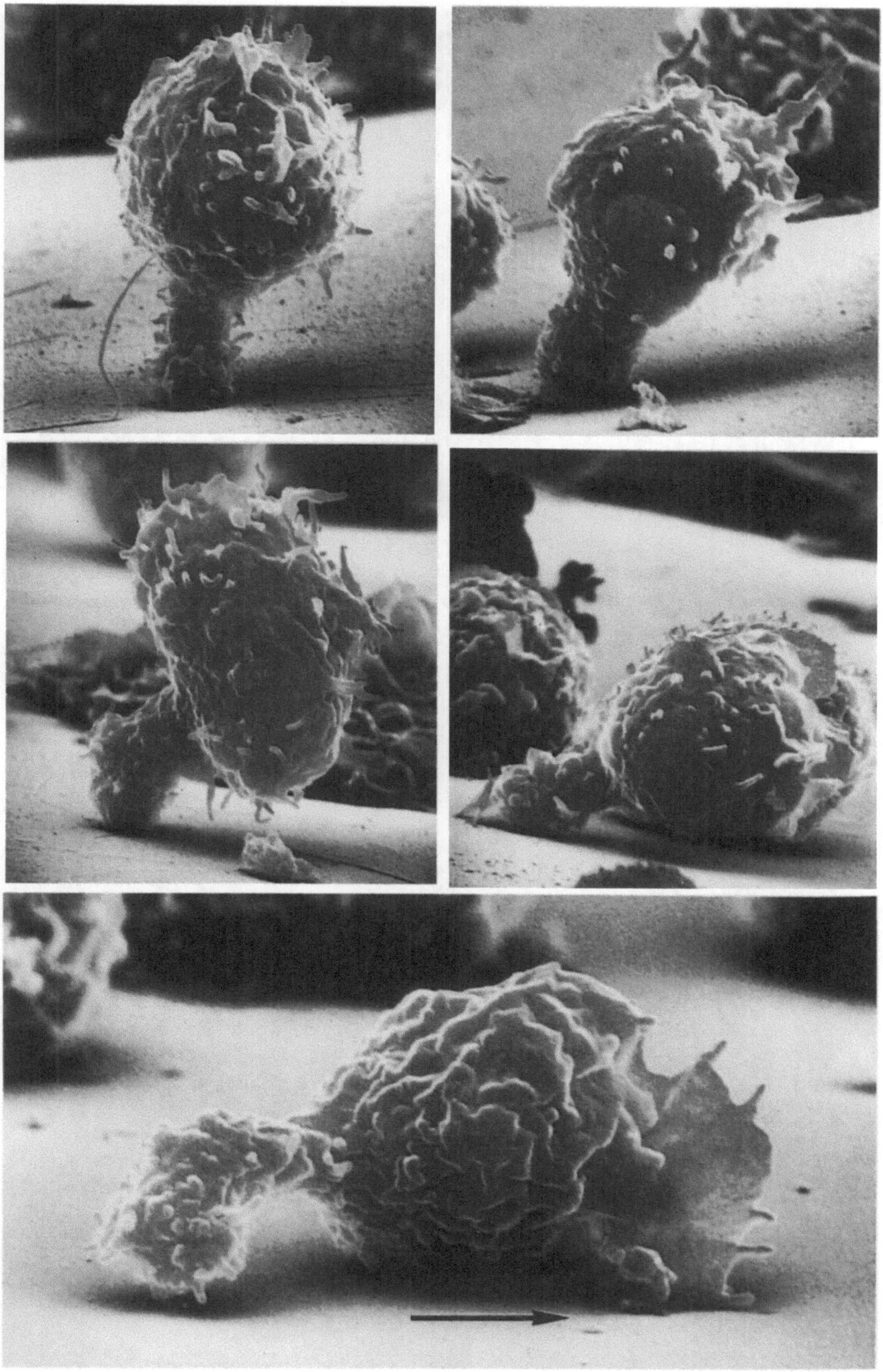

a
b
c
d
e

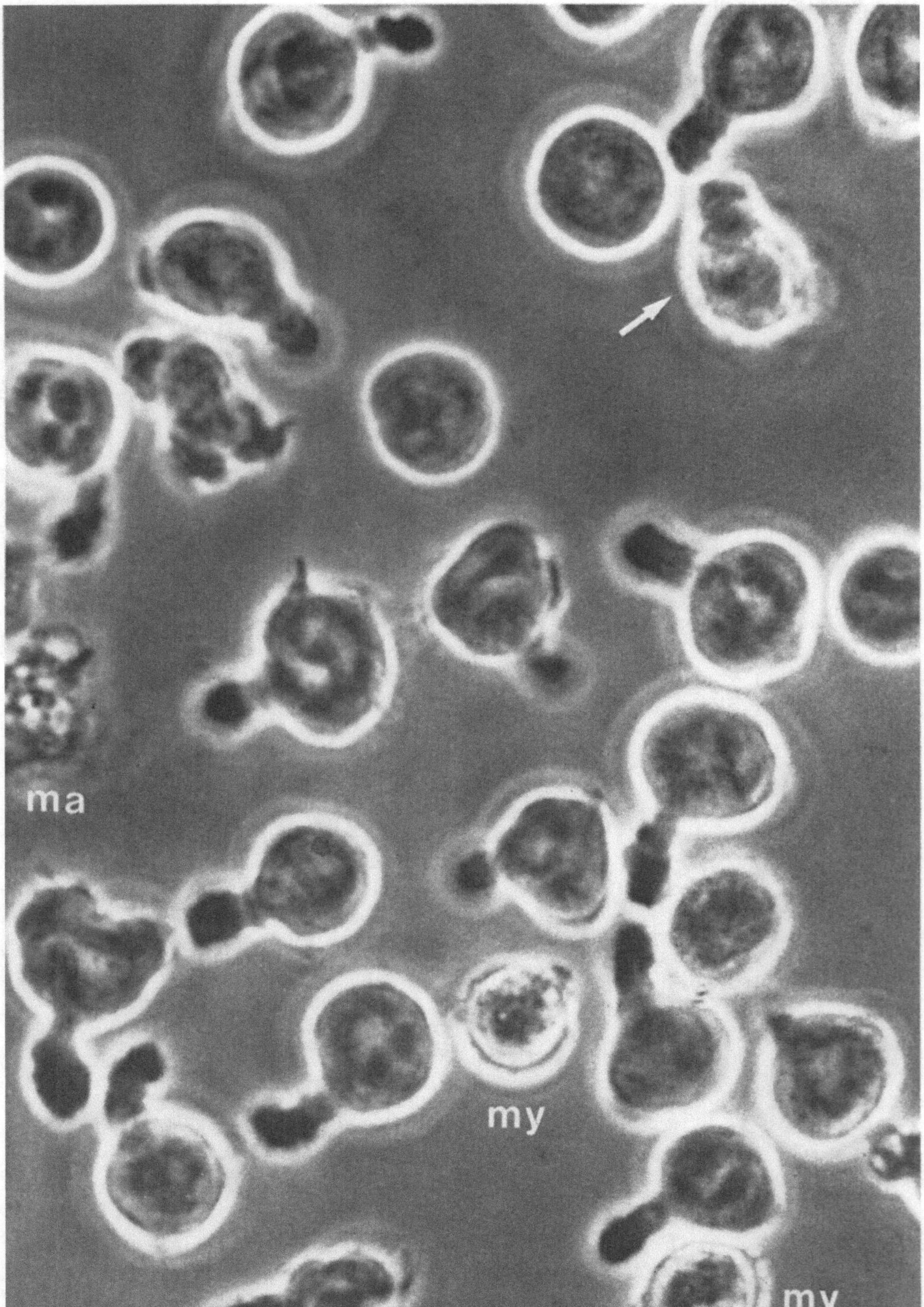
ma
my
my

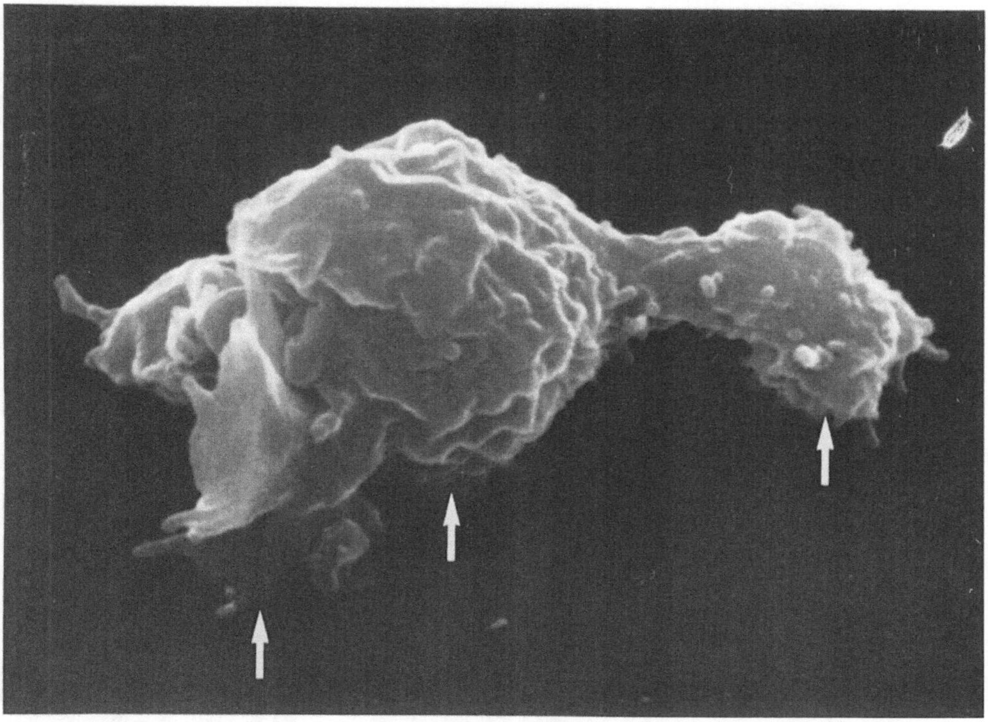

a

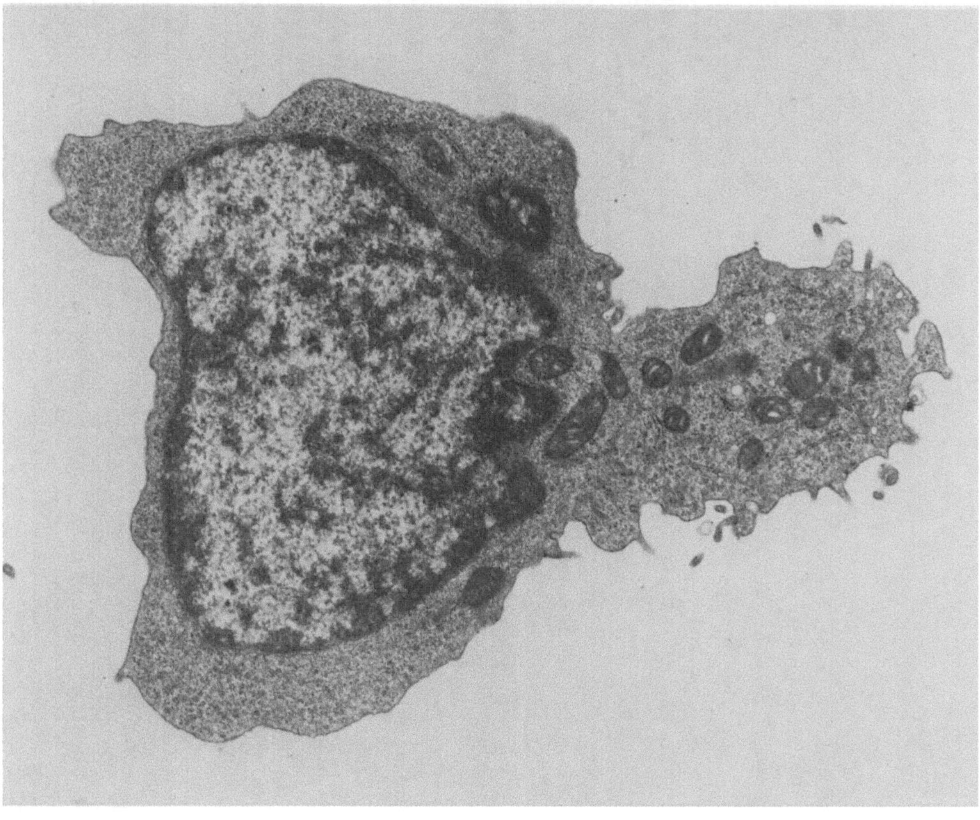

b

17

Figure 10

The phase-contrast photograph from a time-lapse film sequence shows that the polarized configuration of the cells placed on isologous mesentery is the same as for cells adhering to glass. Cinematography further reveals that these cells are locomotive, moving with the same speed as cells on glass

Phase contrast: × *1400*

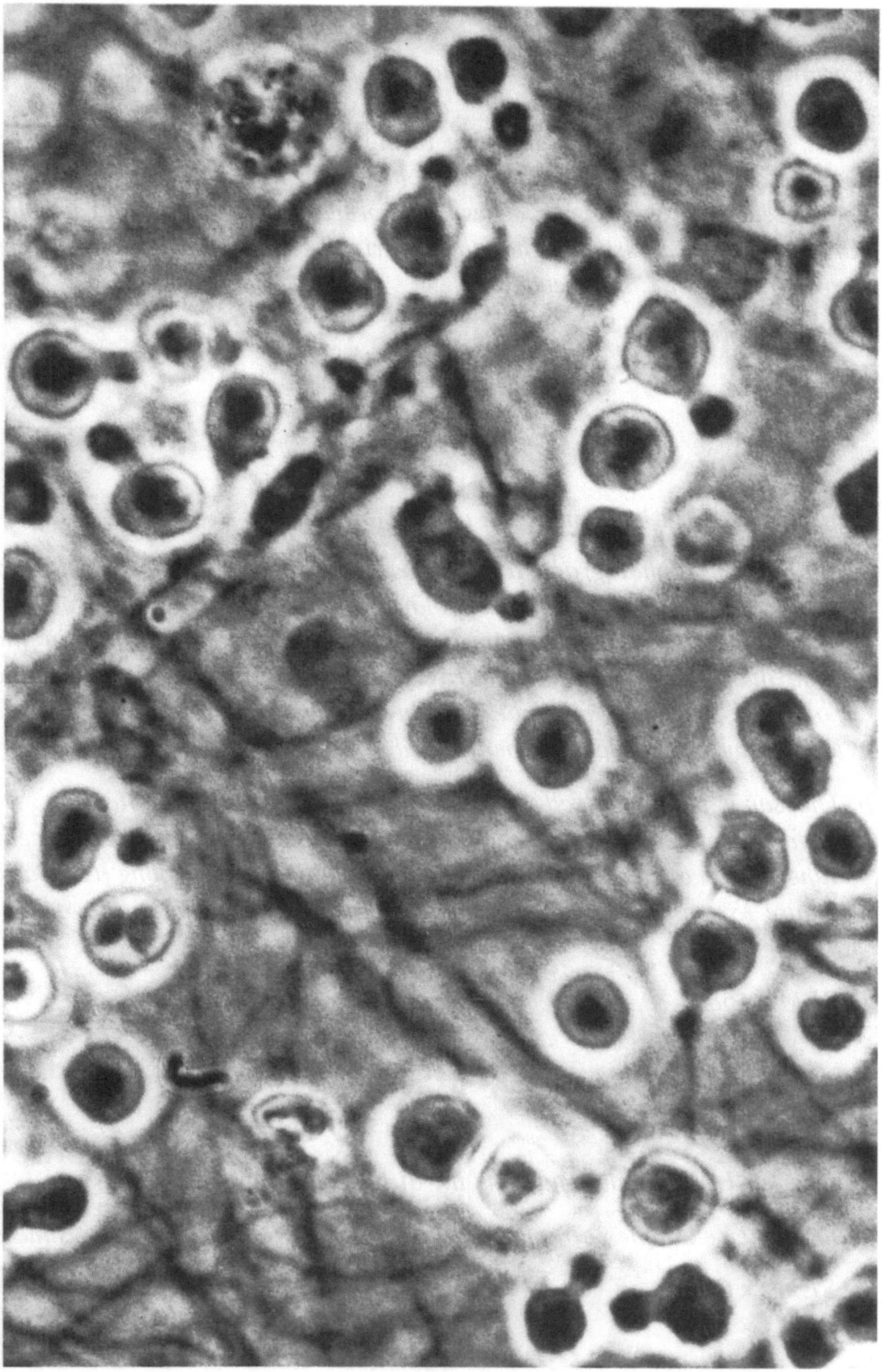

Figure 11. L 5222 leukemia cells in rat bone marrow

In a low-power SEM from a cryofractured fragment of rat bone marrow, various sinuses (s) surrounded by extravascular hemopoietic tissue are shown. A few L 5222 cells, adhering to the luminal surface of the sinuses (arrows) are recognizable by their polarized configuration. As cells in a polarized shape were never observed in preparations of normal rat bone marrow, it can be assumed that the polarized cells in the leukemic bone marrow represent L 5222 cells

SEM: × 1000

Figure 12. L 5222 leukemia cells in rat bone marrow *(see page 22)*

a In this SEM photograph a polarized L 5222 cell can be seen in a sinus (arrow) among other nonpolarized cells and some erythrocytes. The similarity in the configuration of L 5222 cells observed in the bone marrow and under in vitro conditions is evident. Whether the polarized shape of leukemia cells in the bone marrow also indicates locomotive activity cannot be decided

b The same polarized cell is presented in a higher magnification. Its surface architecture is somewhat different from that of L 5222 cells adhering to a glass substrate: the microvilli are very long (see also Figure 13a) and the surface seems to be covered by ridges

a SEM: × 3300
b SEM: × 6800

Figure 13. L 5222 leukemia cells in rat bone marrow *(see page 23)*

Two further examples of polarized leukemia cells in bone marrow sinuses

a SEM: × 6700
b SEM: × 7400

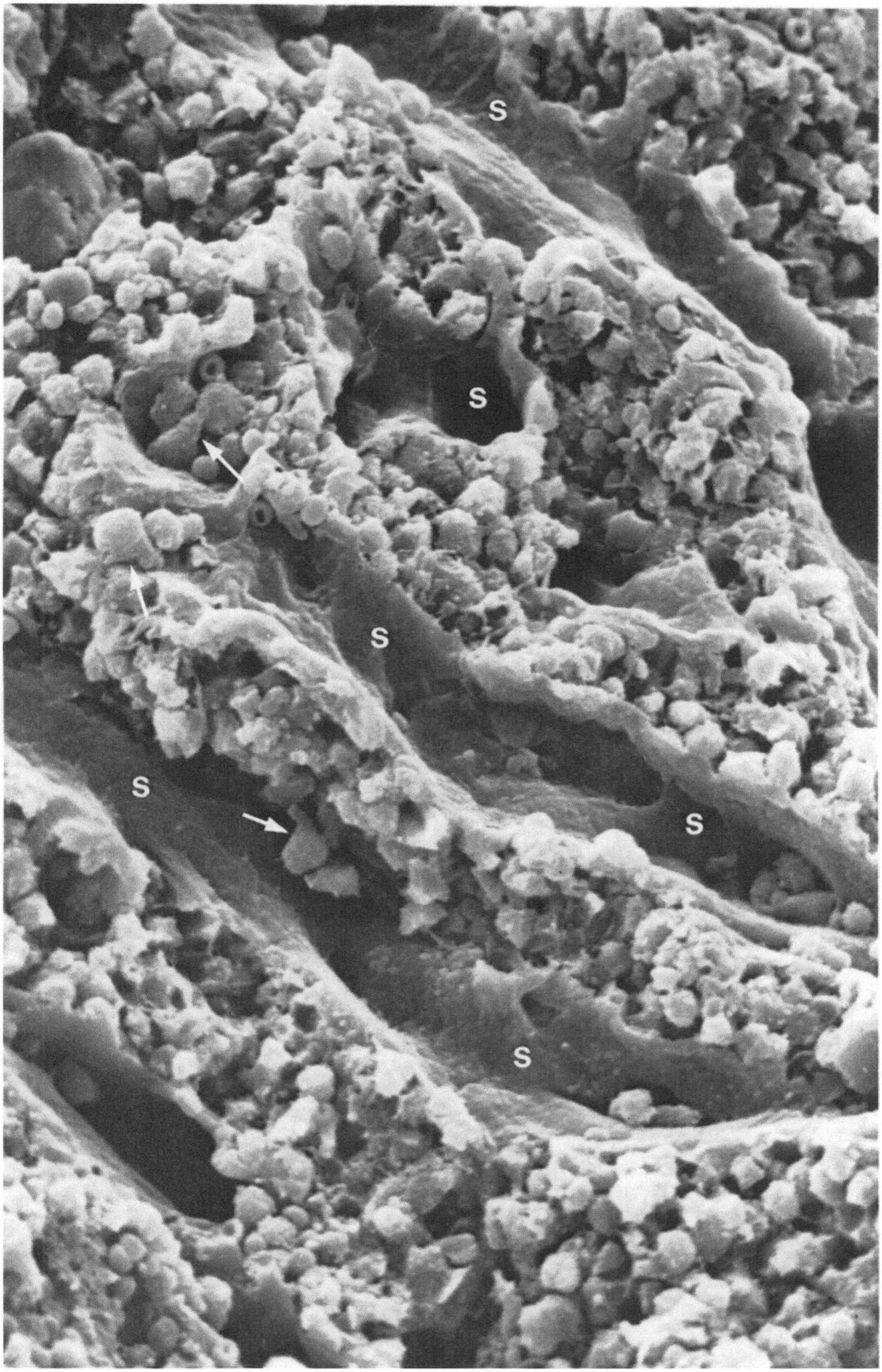
S
S
S
S
S
S

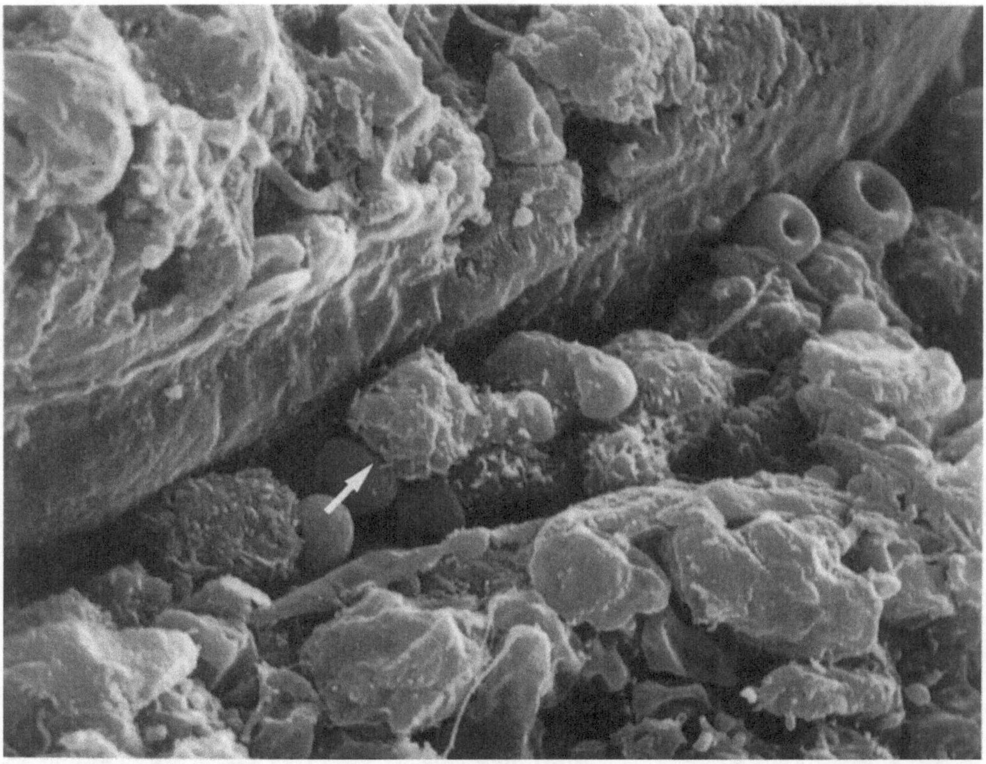

a

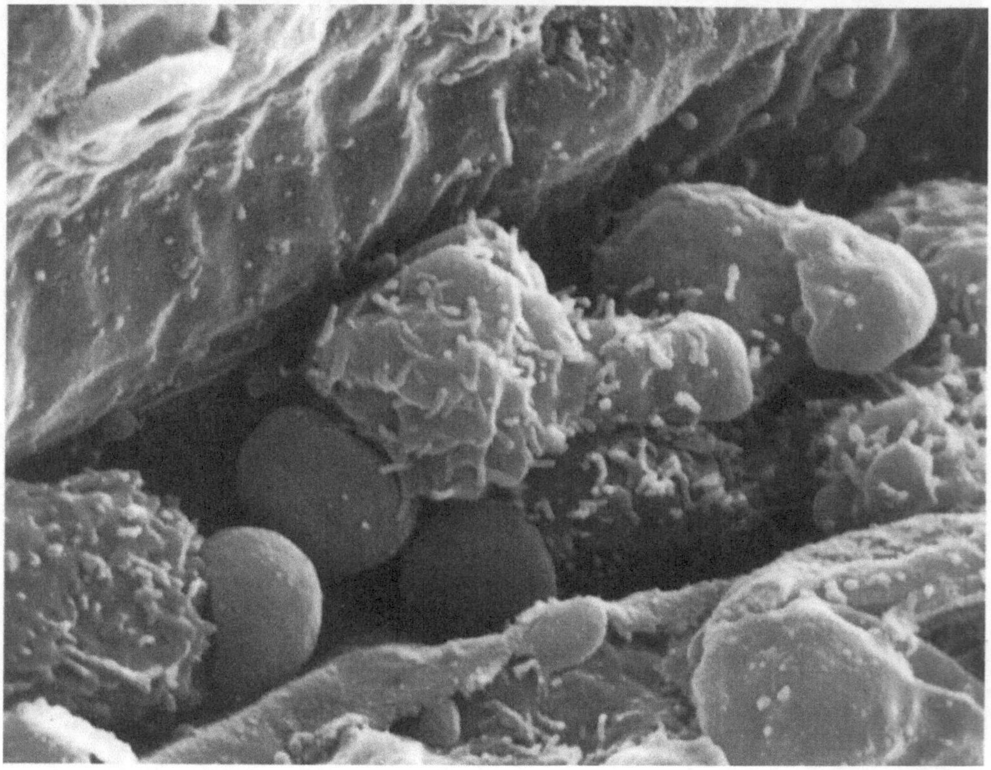

b

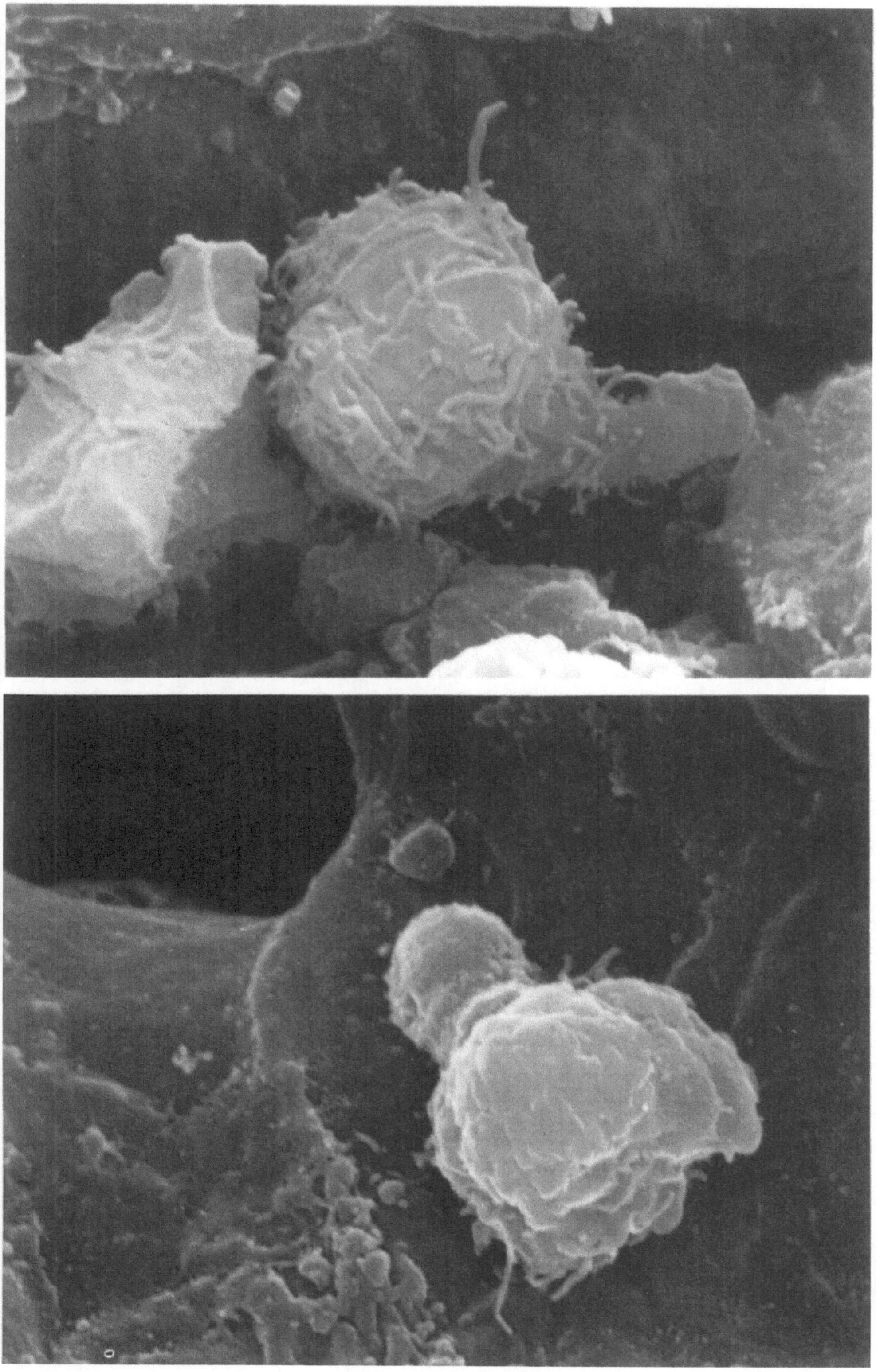

a

b

II. Human Leukemias

Material

Samples of heparinized venous blood from 19 untreated leukemia patients were obtained from the Department of Internal Medicine and the Children's Hospital, both from the University of Zürich. From two adult patients an additional blood specimen was studied during the period of treatment (for technical details see Methodology).

The different leukemia cases are arranged according to the clinical diagnosis: acute myeloid (AML), chronic myeloid (CML), acute myelomonocytic (AMML) leukemias, and acute lymphoid (ALL), chronic lymphoid (CLL), and unclassifiable leukemias. The hematologic data of the individual patients are given in the appendix.

1. Acute Myeloid Leukemias (AML)

This leukemic population, consisting mainly of myeloblasts, exhibits moderate surface motility. Concurrently, the shape of the cells is spheric with the occasional appearance of veil-like cytoplasmic extensions. Slow locomotion occurs, but very infrequently.

Reexamination of the peripheral blood 40 h after therapy had started has given no indication for effects on either shape or behavior of the leukemia cells.

Although cells in the spheric configuration cannot be identified by SEM, they are termed myeloblasts because of their large proportion in the peripheral blood of this patient.

Figure 14. Leukemic myeloblasts (before onset of therapy)

Short microvilli, together with smaller and larger folds and some veils (arrows), cover the surface of the spheric cells. Size, distribution, and number of the surface extensions vary from one cell to the other

SEM: × 9500

Figure 15. Leukemic myeloblasts (before onset of therapy) *(see page 30)*

The large veils shown in the SEM photographs (a and c) indicate surface motility. This could be verified by time-lapse films (b) where similar configurations (arrow) are the result of this activity. Several cells have settled on a spread thrombocyte (a)

a SEM: × 7800 b Phase contrast: × 1900 c SEM: × 9300

Figure 16. Leukemic myeloblasts (before onset of therapy) *(see page 31)*

a Spheric cells are settled on a spread thrombocyte. Again the difference in surface architecture of individual cells is evident

b Spheric cell settled on the glass substrate by means of a cytoplasmic attachment plate

a SEM: × 4700 b SEM: × 16,000

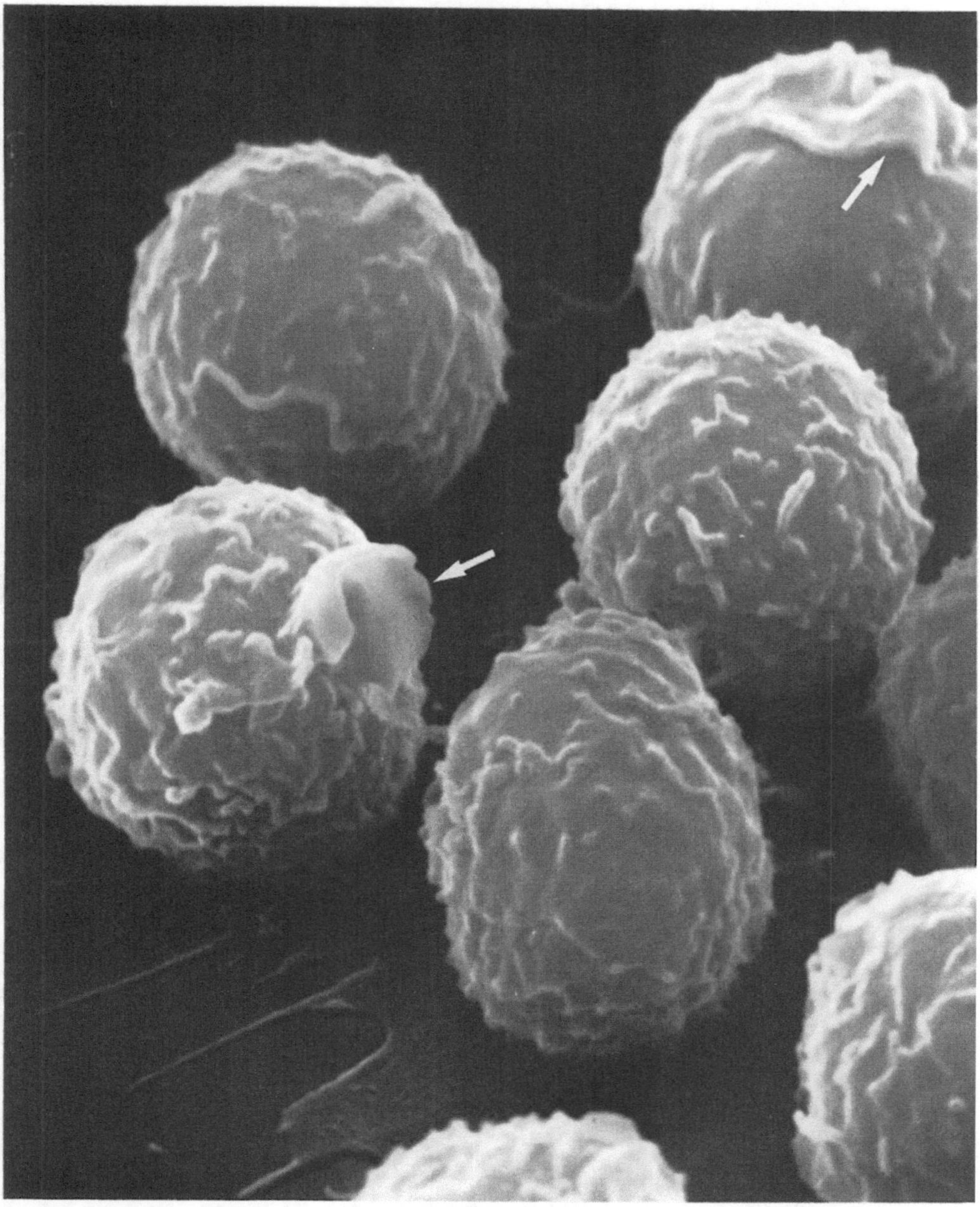

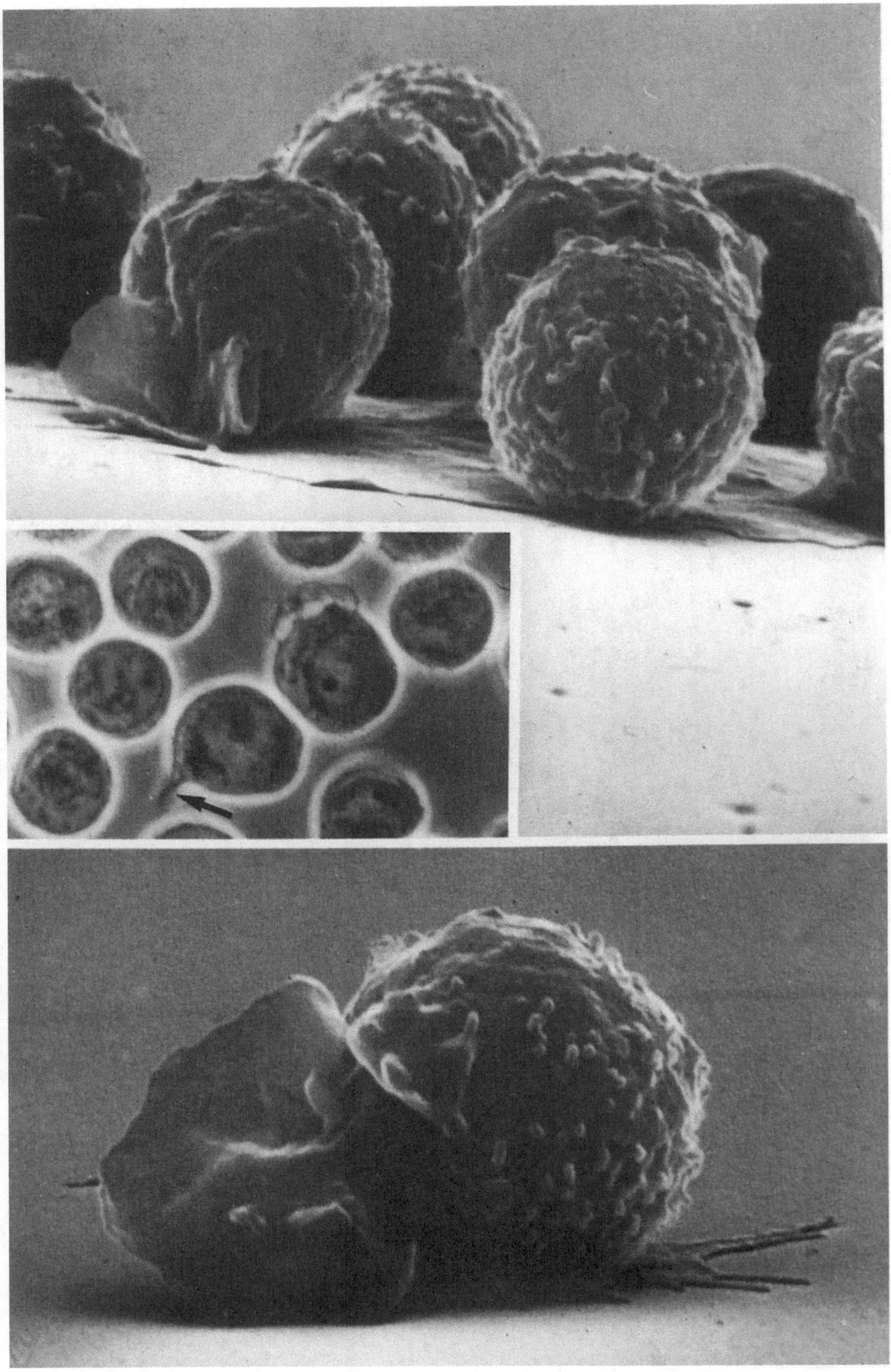

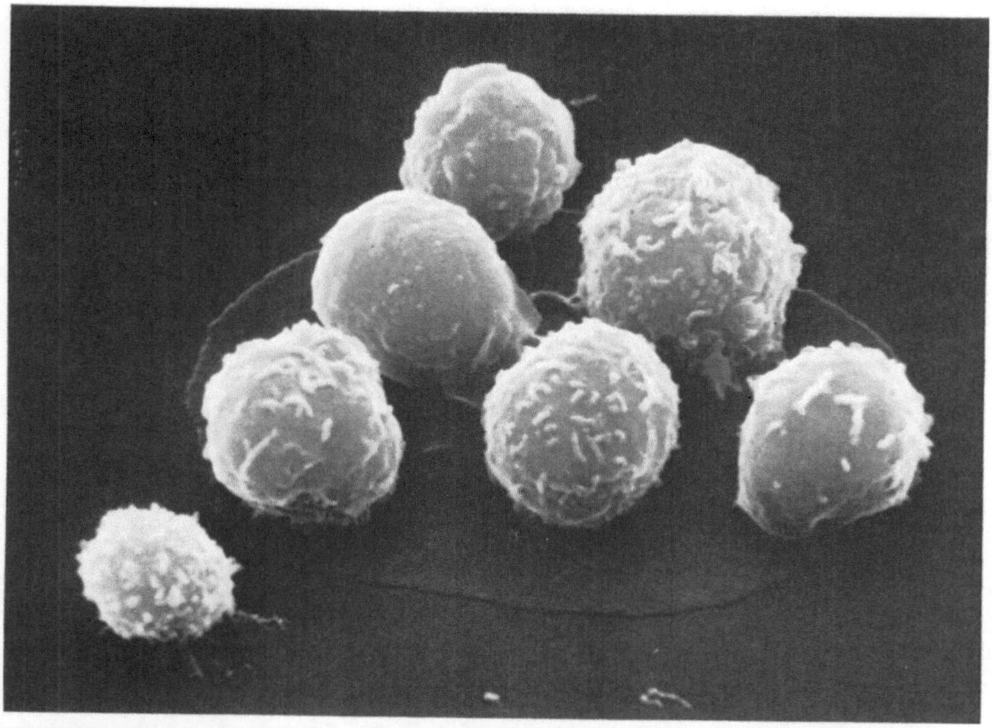

a

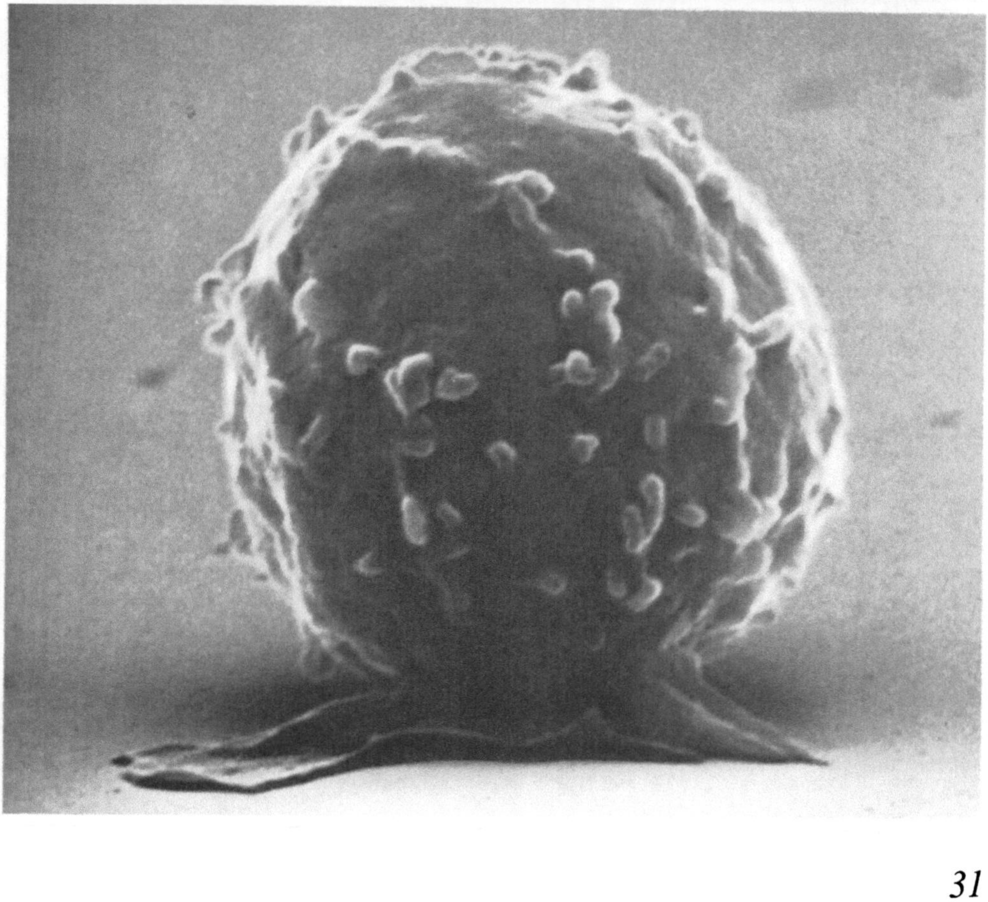

b

Figure 17. Leukemic myeloblasts (during therapy)

Neither the diversity in surface architecture of the spheric cells (a) nor surface motility are changed since the onset of therapy. Note that the preference for settling on spread thrombocytes is also maintained

a SEM: × 6300
b Phase contrast: × 2600

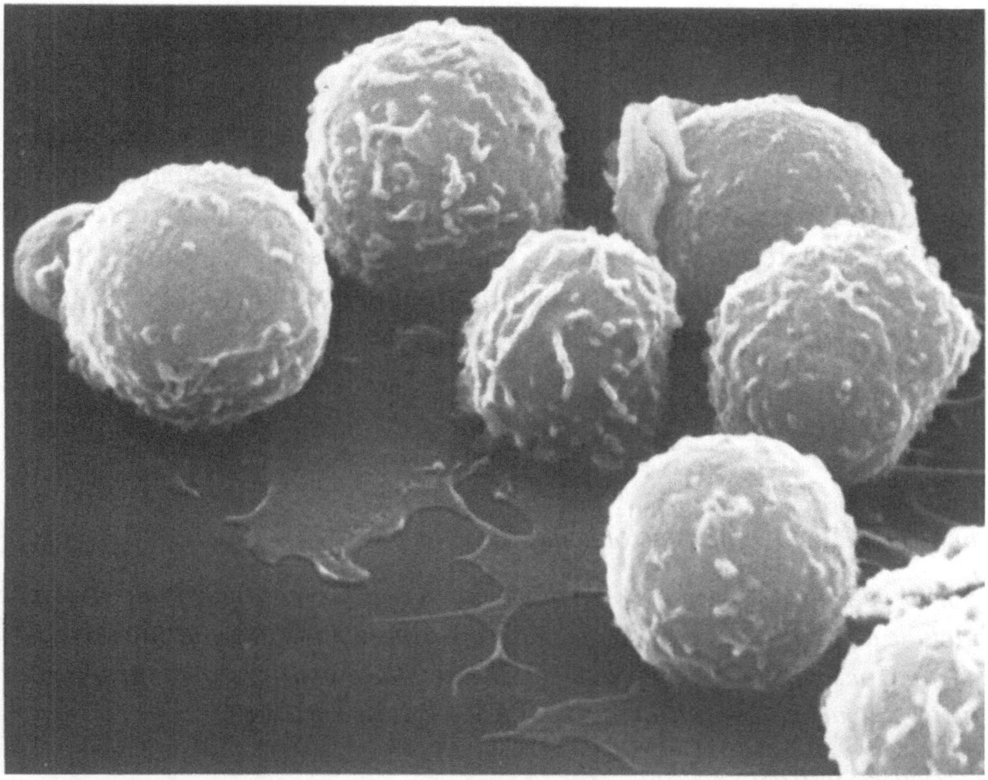

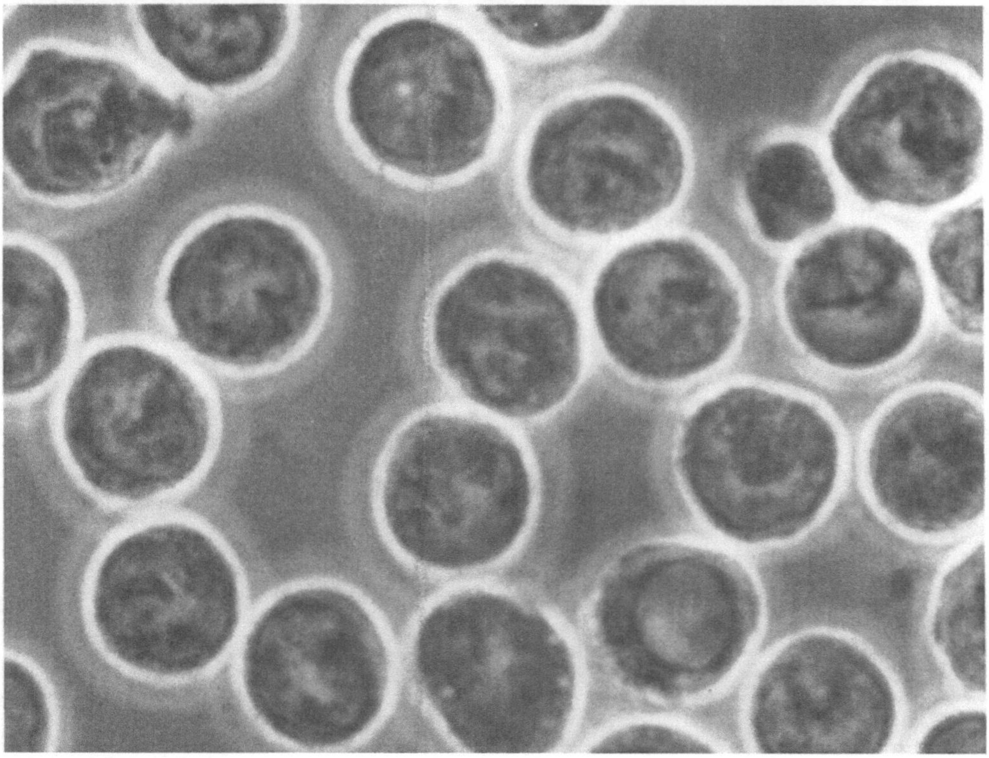

This leukemia also consists mainly of myeloblasts which, again, exhibit no locomotion. However, their surface motility, if compared to the cells of the first case, is increased and the surface architecture, accordingly, more heterogenous. Occasionally, the cells even assume a polarized configuration. Because of their predominance in the leukemic population, spheric cells are again termed myeloblasts.

Figure 18. Leukemic myeloblasts

Some cells are spheric, some have an irregular configuration that can include a partly polarized shape. These features, indicative of the increased surface activity, can be recognized in the phase-contrast photograph from a film sequence as well as in the TEM picture (arrows)

a Phase contrast: × 1200
b TEM: × 4200

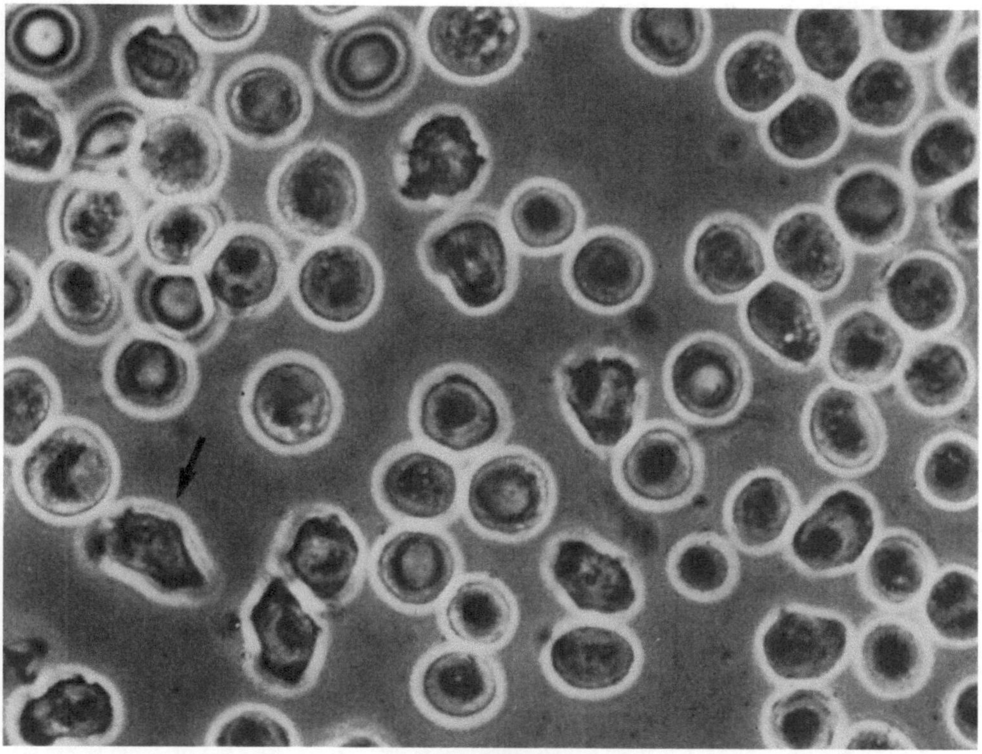

a

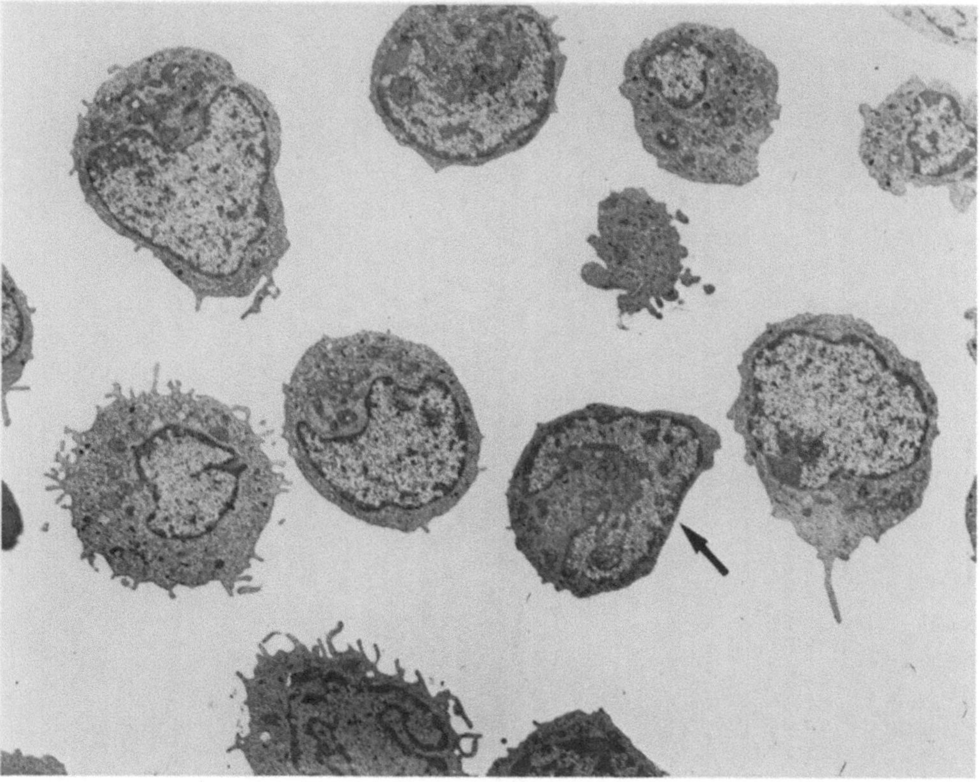

b

Figure 19. Leukemic myeloblasts

a *While one cell is spheric with a fairly regular surface configuration, another, still round, shows small folds and veils. A similar picture is presented by a third cell that is slightly polarized by the extension of a short foot fastened to the substrate by means of attachment filaments*

b *Folds and veils are suggestive of surface motility*

a SEM: × *7300*
b SEM: × *12,700*

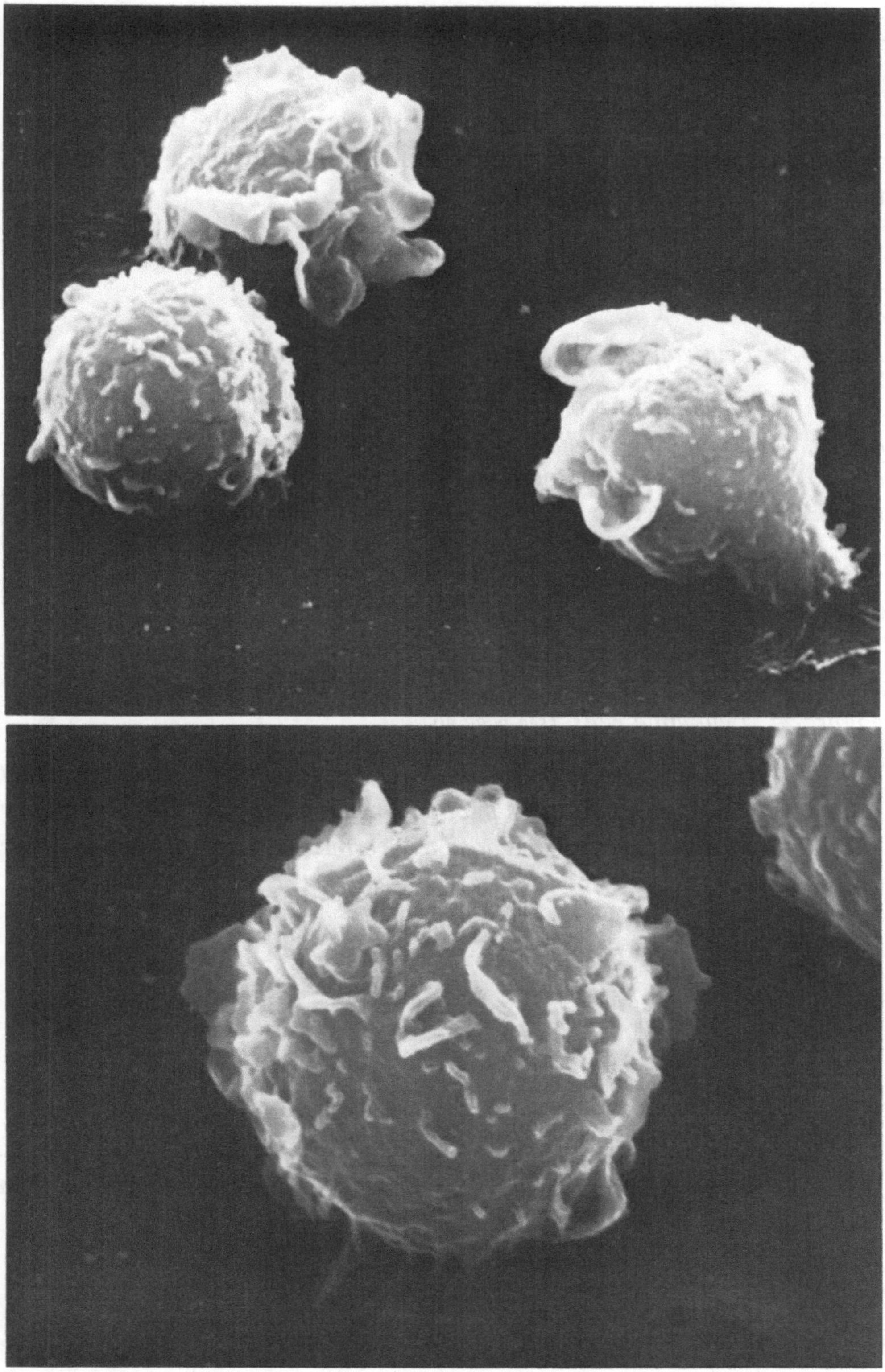

This population consists of small-sized spheric cells with surface motility as the main activity. Few cells only show on-spot motility and locomotion.

Figure 20. Leukemic population

The surface architecture of the different cells varies: there are microvilli, ridges, and folds. The polarized cell is either locomotive or engaged in on-spot motility. The photograph does not allow this distinction to be made. There are several thrombocytes (th), spread and spheric

SEM: × 8400

Figure 21. Leukemic cells *(see page 40)*

The difference in surface extensions of cells from the same population is shown for two representatives in a higher magnification. This divergency together with the spheric shape does not permit the identification of these cells

a SEM: × 17,300
b SEM: × 15,300

Figure 22. Leukemic myeloblasts *(see page 41)*

The cell in the SE micrograph (a) has the typical configuration of a blast cell. This is verified by TEM (b)

a SEM: × 16,000
b SEM: × 12,000

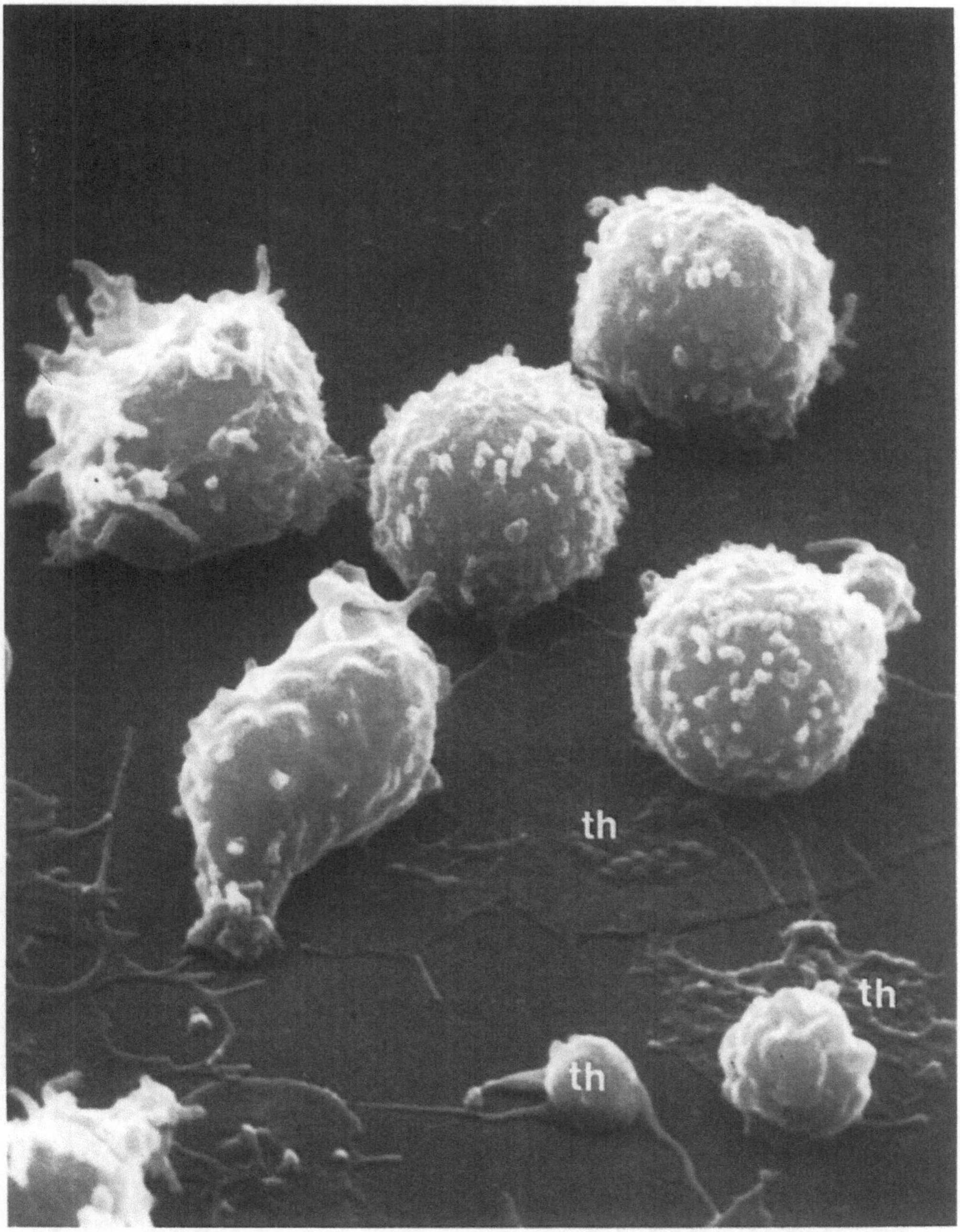
th
th
th

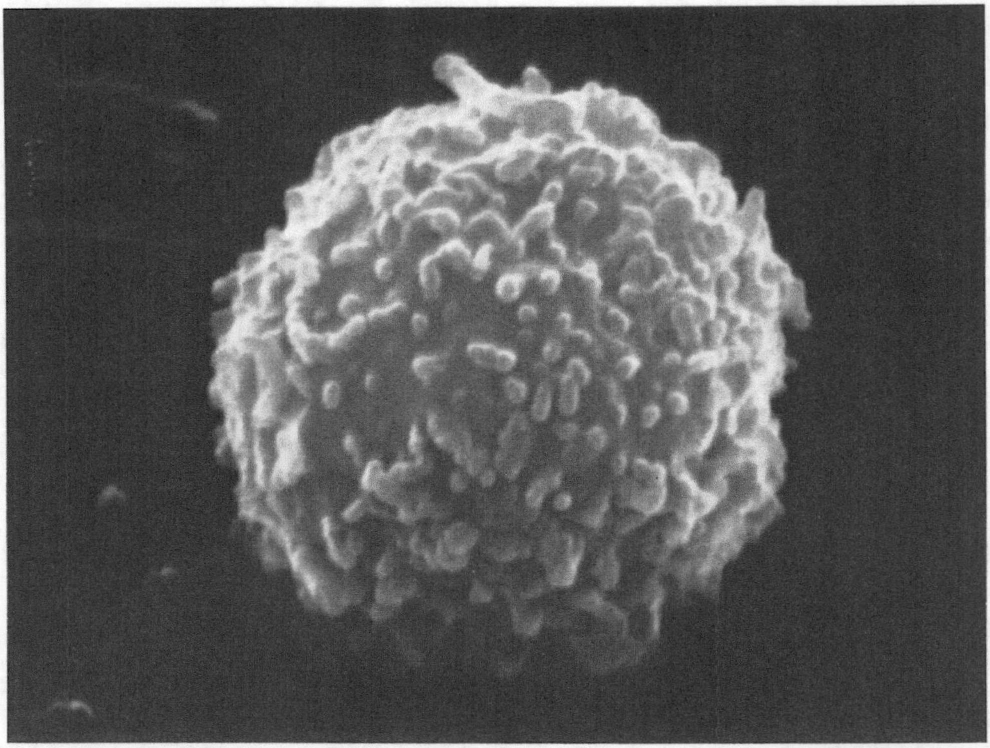

a

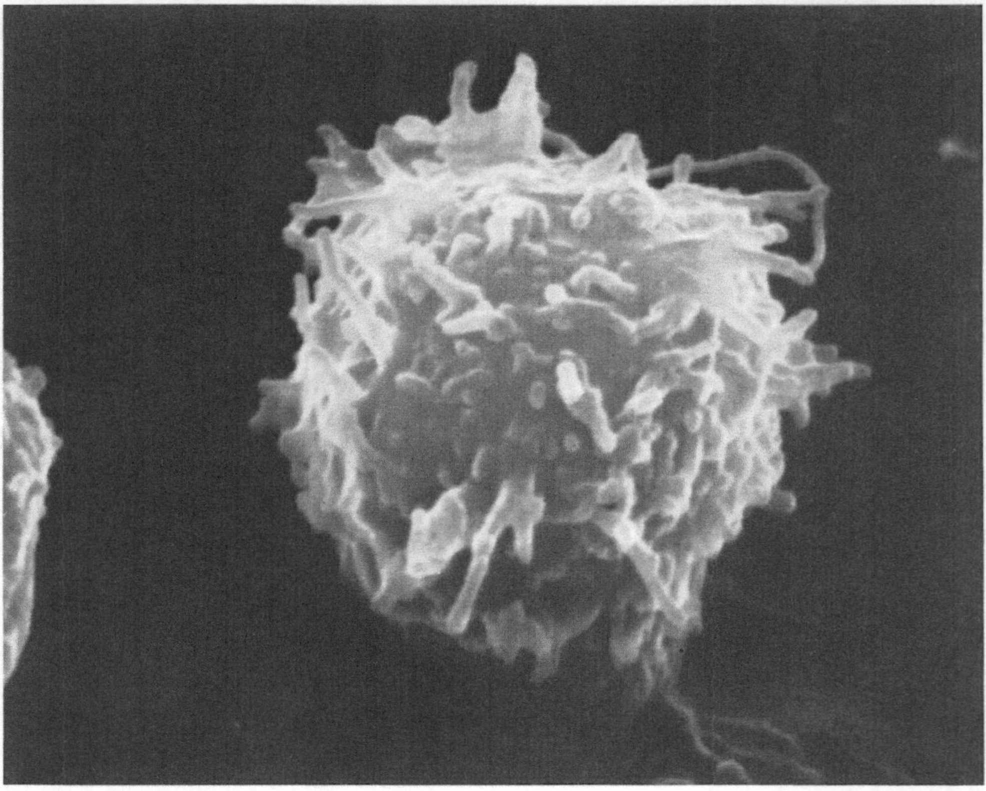

b

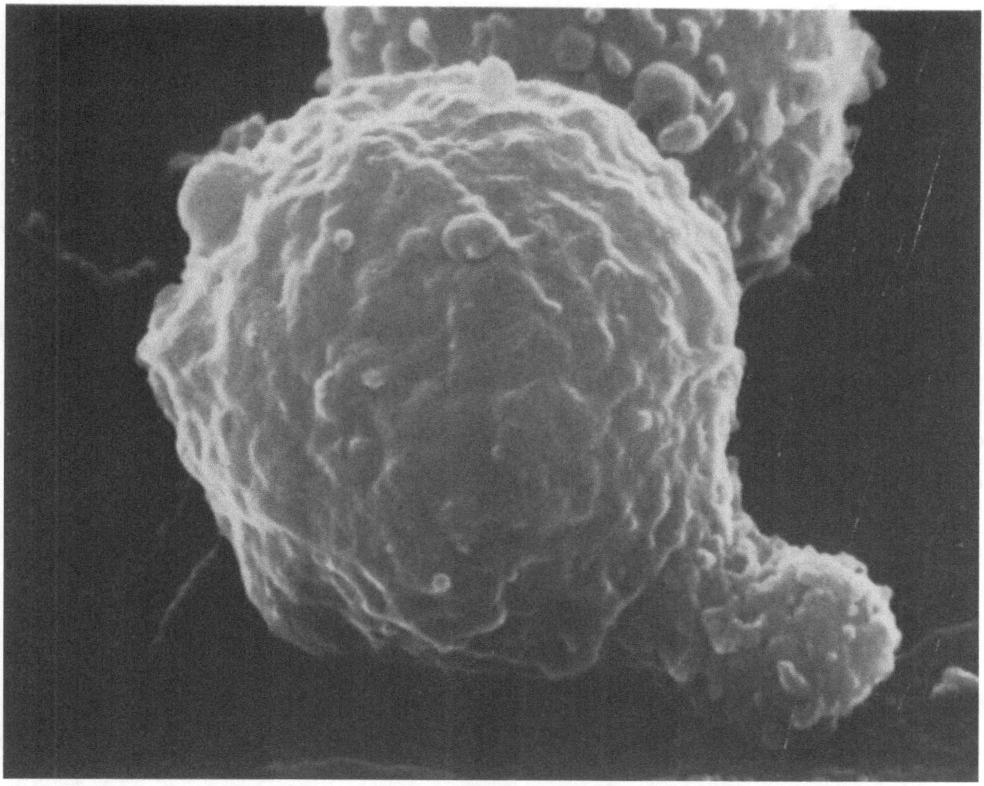

a

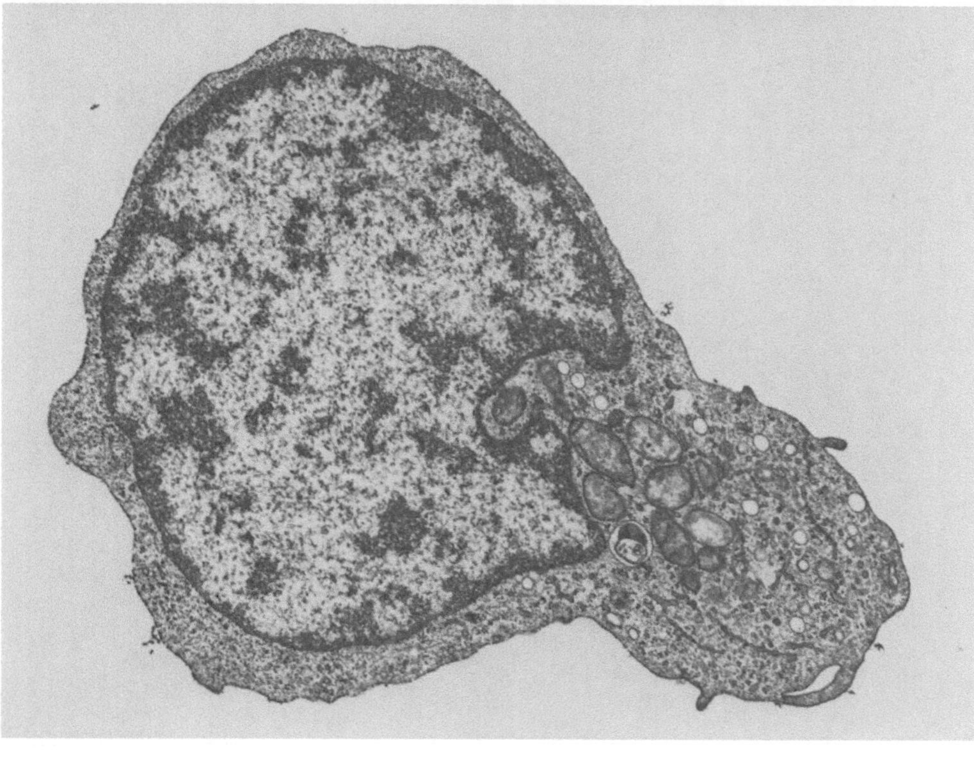

b

Figure 23. Leukemic myeloblasts

a *The three phase-contrast pictures from a time-lapse sequence show a myeloblast during on-spot motility. The cell is fastened to the glass substrate by an extension that is used as a stalk for the performance of oscillating movements. The cell in the first position appears blurred because of its swaying out of focus. Note the changes of shape that the cell undergoes during its activity*

b *The upright position of the polarized cell (arrow), fastened to a spread thrombocyte, suggests on-spot motility*

a *Phase contrast:* $\times$ *1900*
b *SEM:* $\times$ *8200*

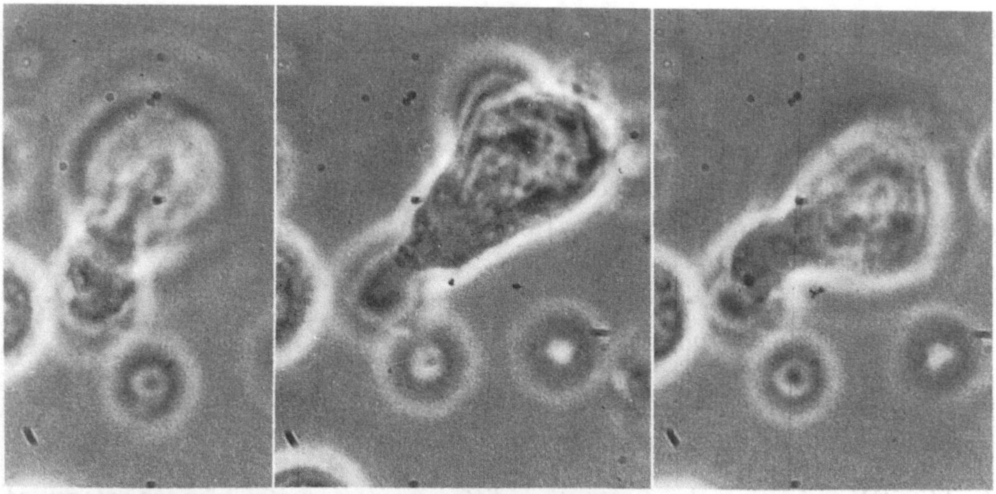

a

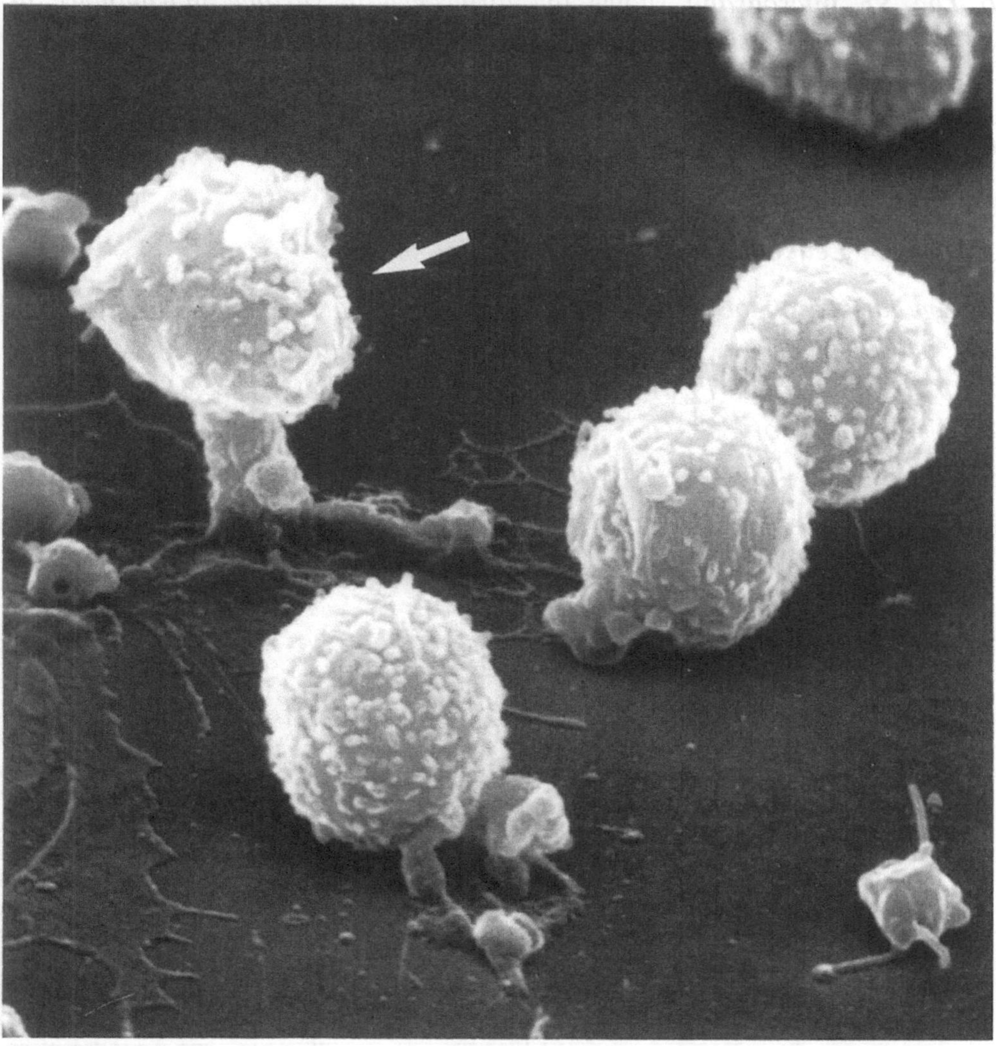

b

A different situation is presented by this population of leukemic myeloblasts: many cells are engaged in slow locomotion moving in the polarized configuration of the blast pattern of locomotion.

Figure 24. Leukemic population

a The prevailing polarized configuration of the locomotive cells is well documented by the phase contrast photograph from a film sequence

b The SE micrograph shows the difference in surface morphology of both spheric and polarized cells that have microvilli, blebs, ridges, and folds

a Phase contrast: × 2200 b SEM: × 8300

Figure 25. Leukemic myeloblasts *(see page 46)*

The polarized shape with a rounded anterior part and a well-developed tail is characteristic for blast cells. This configuration is preserved by both SEM (a) and TEM (b). Many contact zones, seen especially well at a tilt angle of 85° (a), suggest that this cell has been locomotive at the time of fixation. The arrow indicates the direction of movement

a SEM: × 15,600 b TEM: × 10,200

Figure 26. Leukemic myeloblasts *(see page 47)*

a The polarized configuration of this myeloblast suggests locomotive activity

b The drawing, prepared from a time-lapse film sequence, presents the pathways of two locomotive blast cells that moved slowly at a rate of 2.3 and 3.2 µm/min. The cells covered the distance of ca 30 µm in 13 and 10 min respectively

a SEM: × 16,400 b Drawing from a time-lapse film sequence

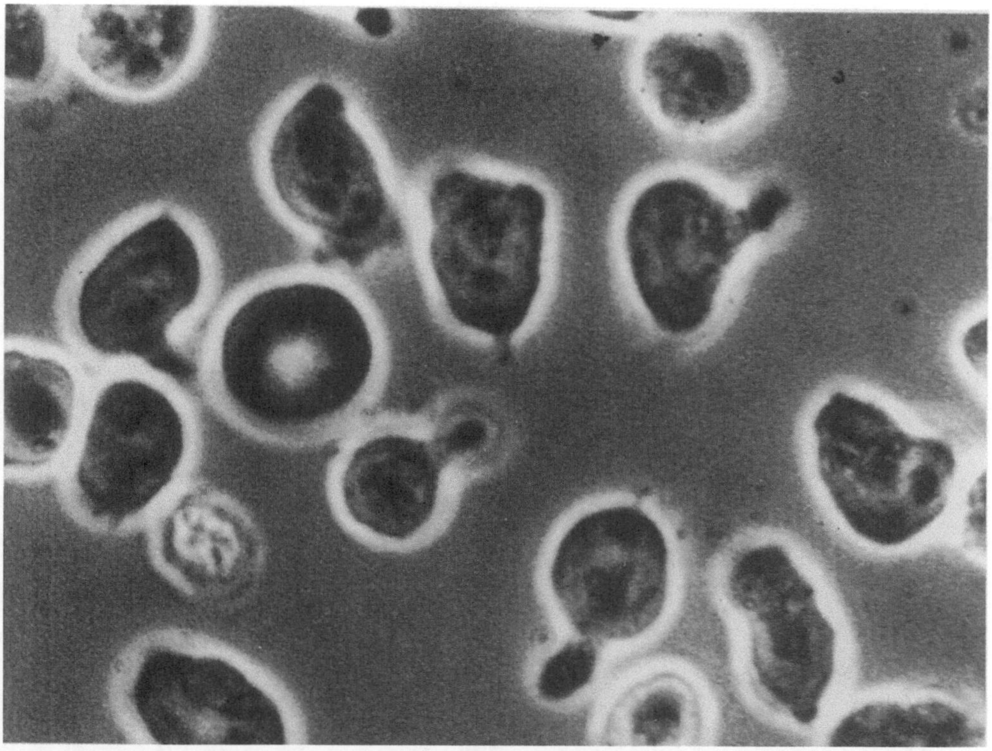
a

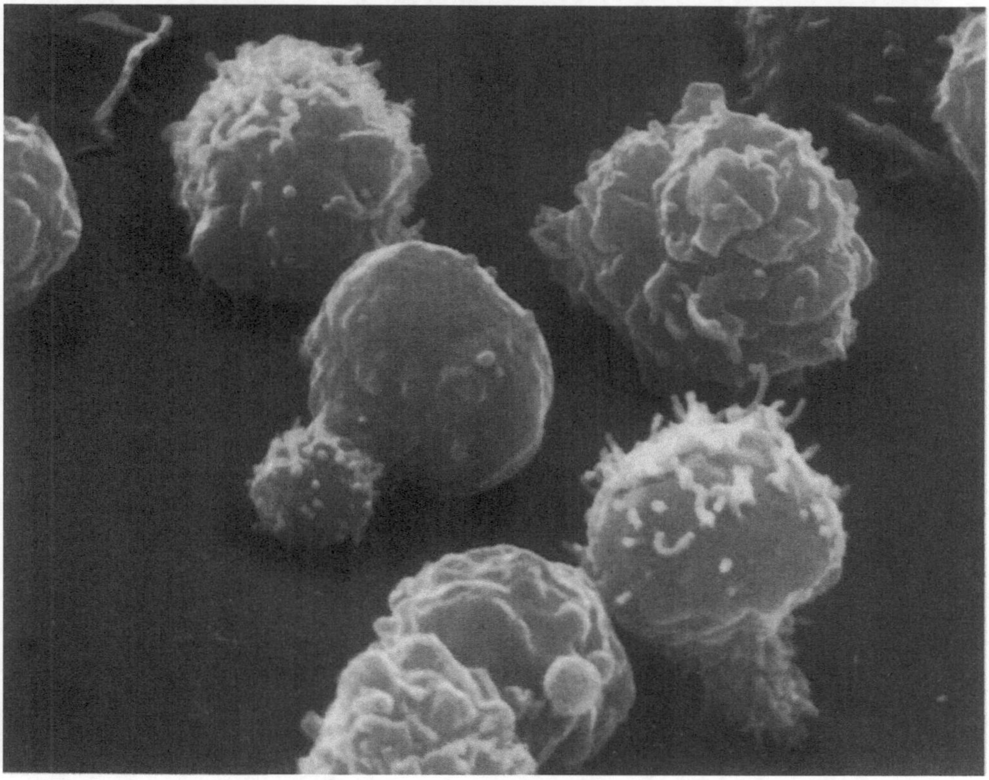
b

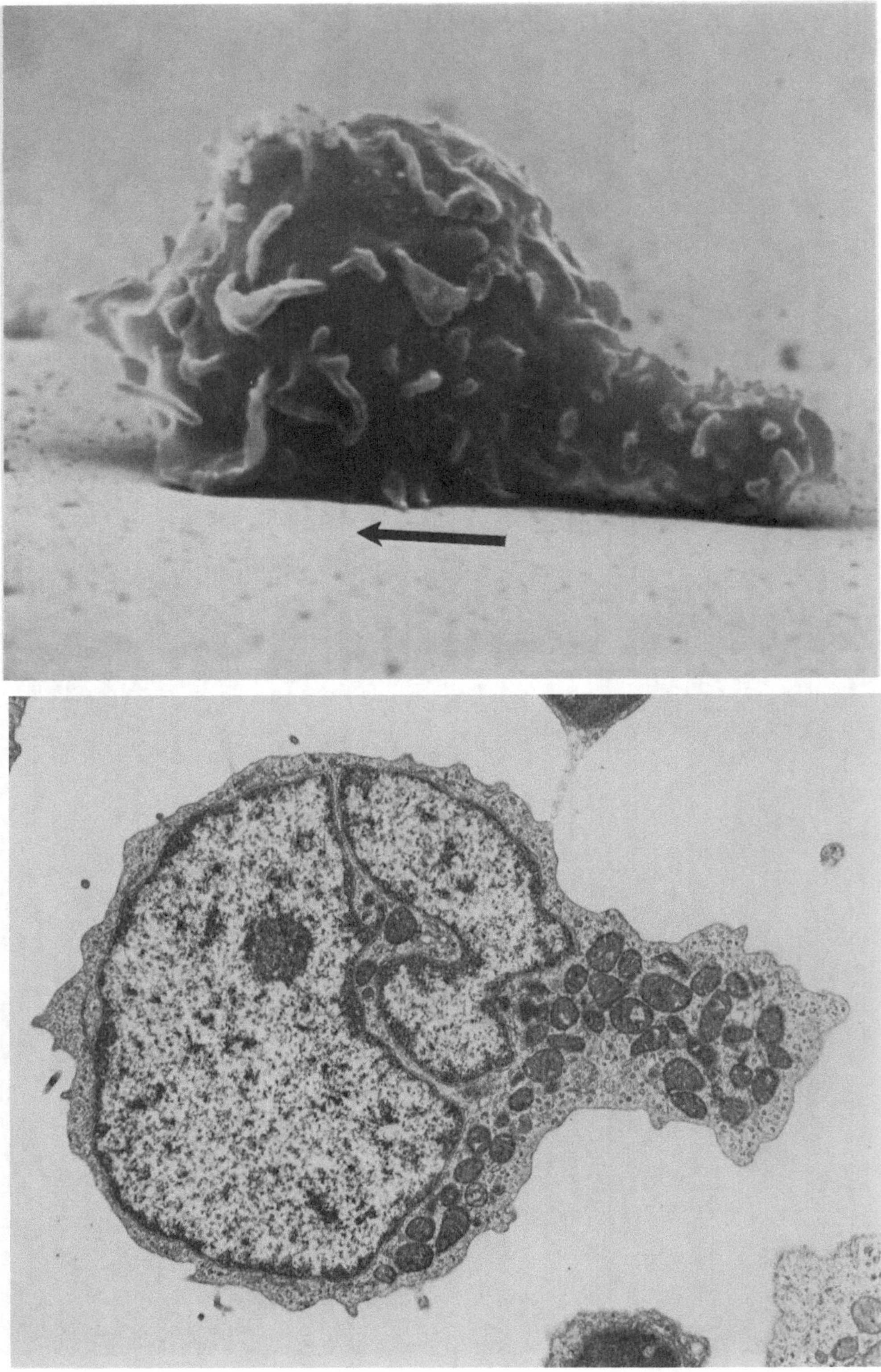
a
b

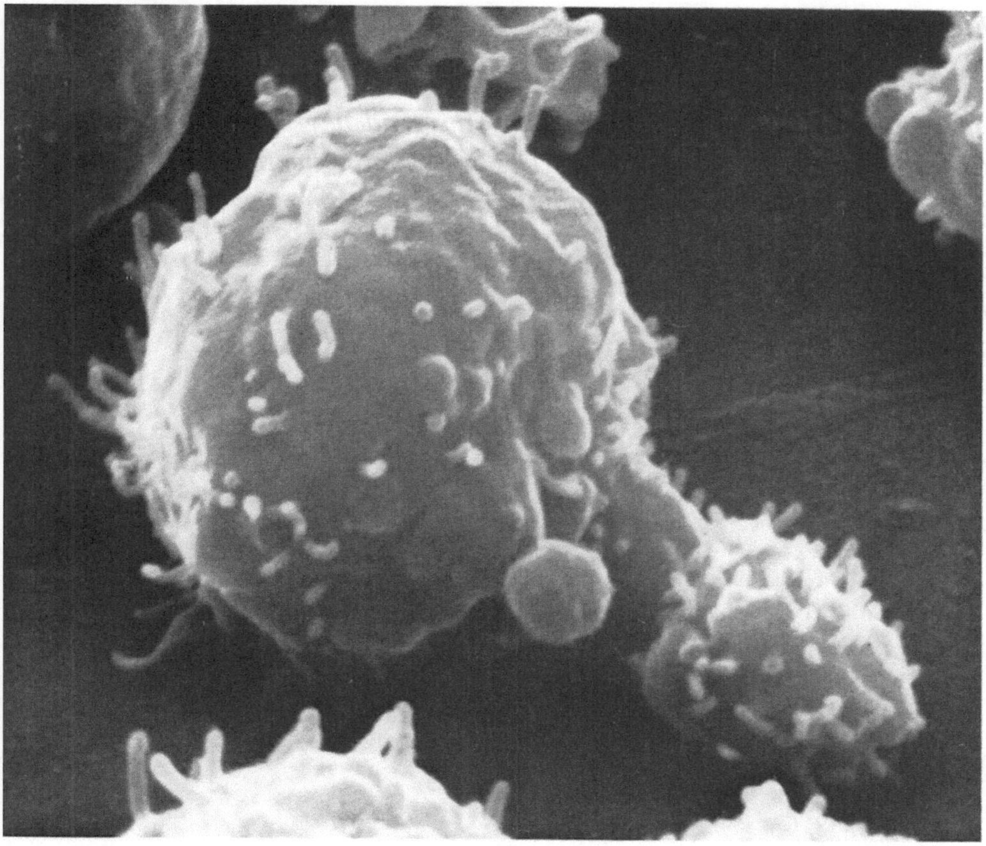

a

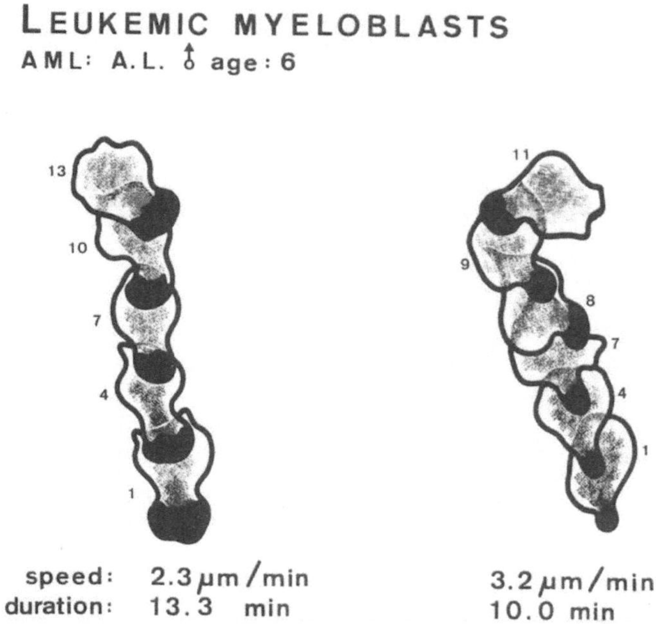
LEUKEMIC MYELOBLASTS
AML: A.L. ♂ age: 6
13
10
7
4
1
11
9
8
7
4
1
speed: 2.3 µm/min
duration: 13.3 min
3.2 µm/min
10.0 min

b

Figure 27. Leukemic myeloblasts

A special feature of the polarized myeloblasts of this population is shown in the two SEM photographs (a, b): the tendency to make contact with the front while the posterior part is lifted into the medium. Contact is established by broad folds. Similar folds cover the rest of the cell body

a SEM: × 14,200
b SEM: × 14,600

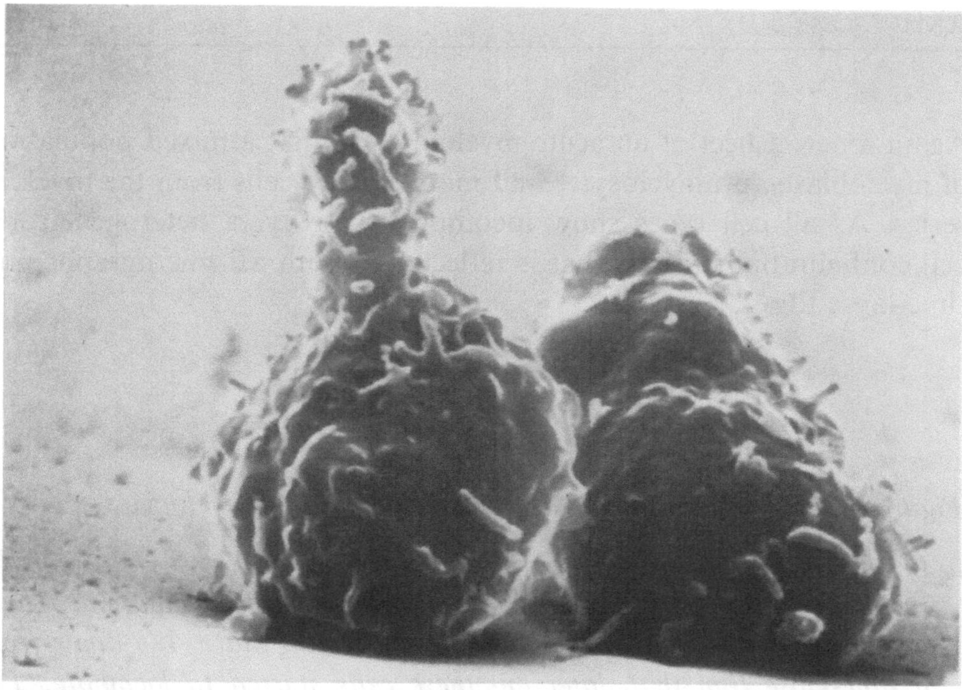
a
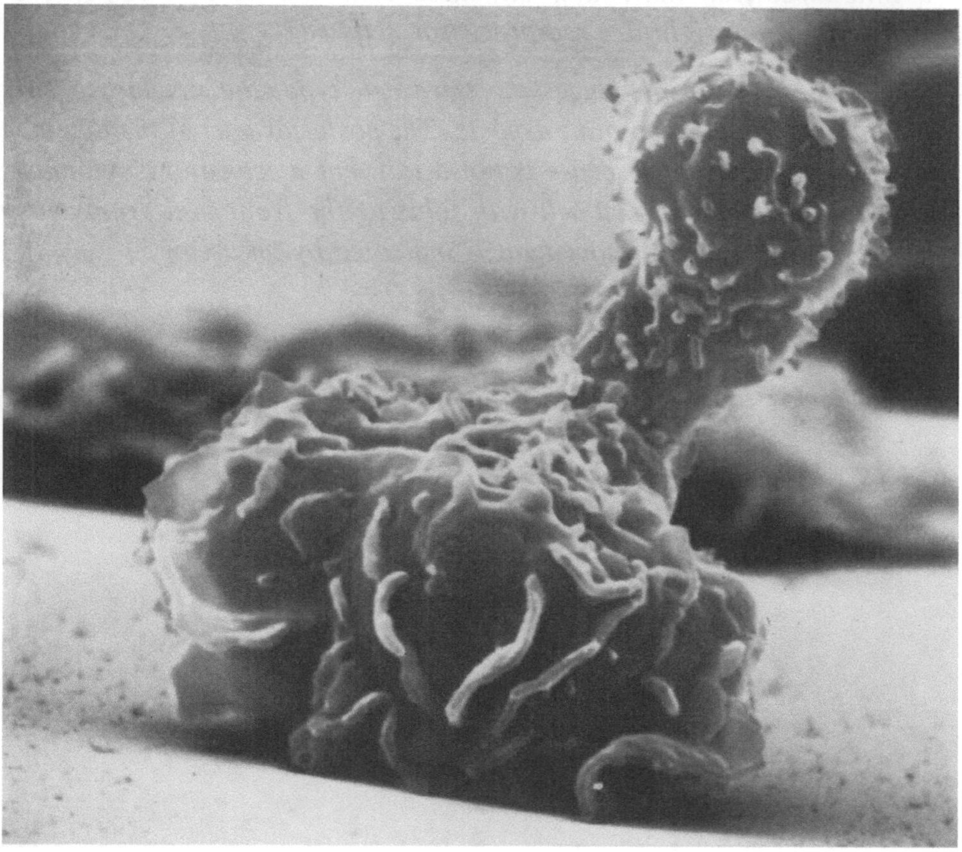
b

Again a new aspect of an acute myeloid leukemia: a mixed population of myeloblasts, promyelocytes and more mature cells from the myeloid series. As all cell types show locomotive activity, a heterogeneity in cell configurations results that is reflected in both SE micrographs and time-lapse films.

Figure 28. Leukemic myeloblasts, promyelocytes and myelocytes

a In the phase-contrast photograph, several myeloblasts (mb) can be recognized, two in the polarized configuration of locomotion. There are also some promyelocytes (pm) in the resting state. We know from a time-lapse film that, later on, these cells started to locomote. The two cells to the right are red precursor cells (e)

b The SEM picture introduces the third cell type, the myelocyte (my), caught in various locomotive aspects. The posterior part of the myelocyte in the upper left hand corner is not a tail, but a transitory cytoplasmic extension. The elongated cell with folds at the front is a promyelocyte (pm), its direction of movement is indicated by an arrow

a Phase contrast: × 2200
b SEM: × 4300

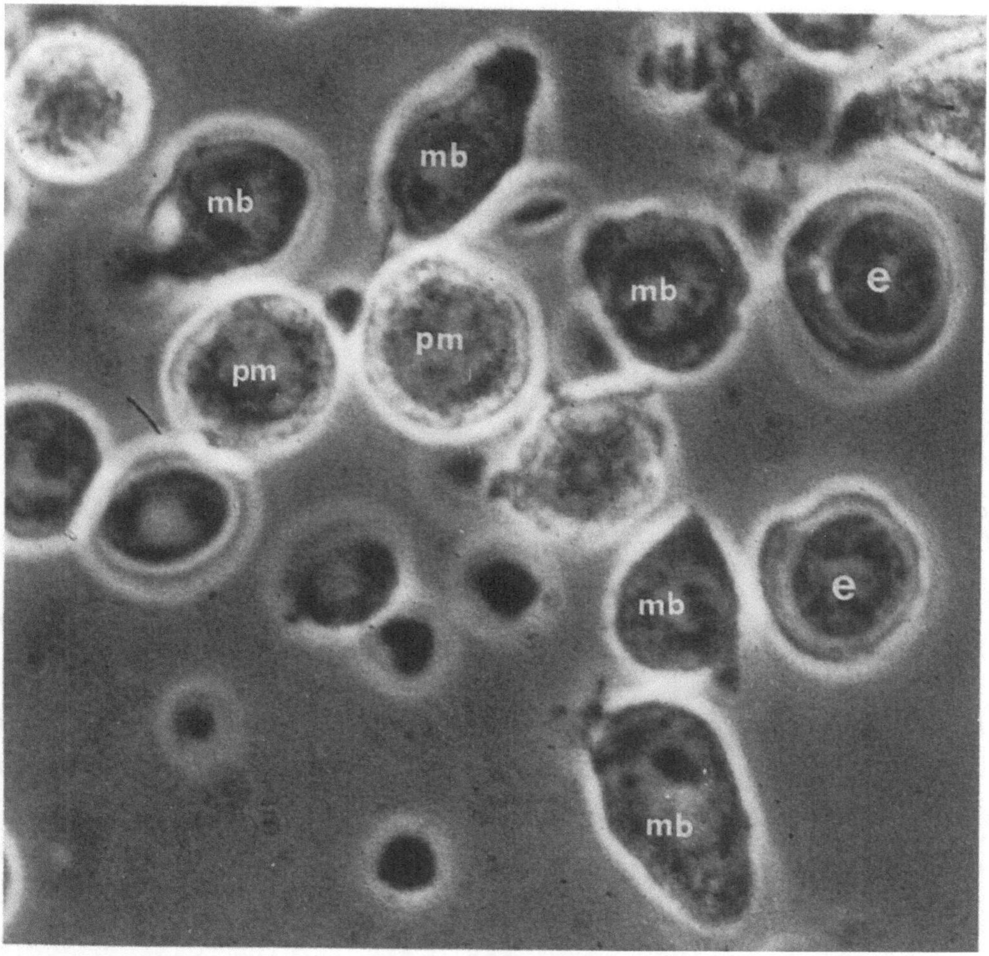

a

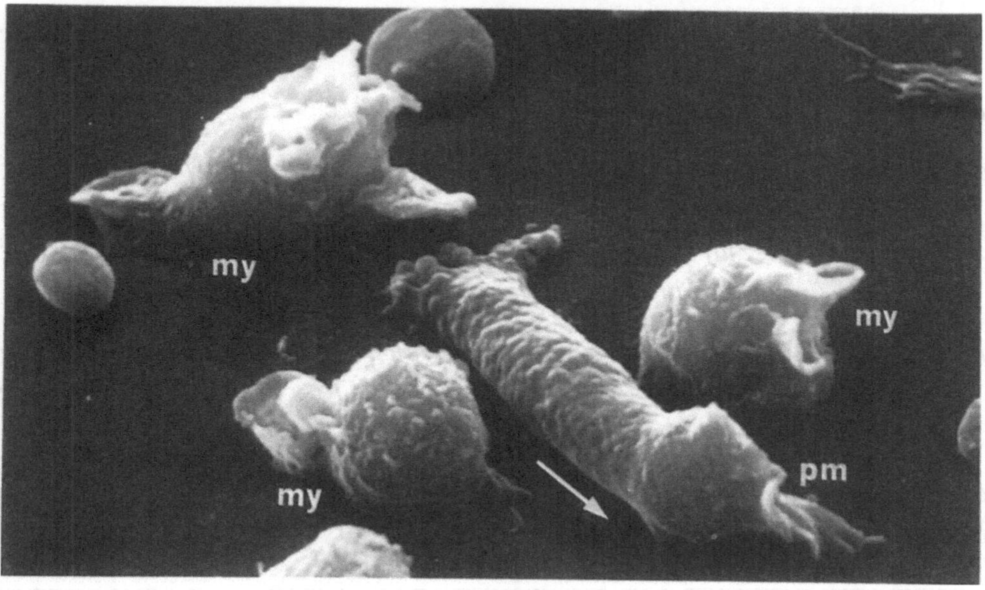

b

Figure 29. Leukemic myeloblasts

These cells show the polarized configuration typical for blast cells. The surface architecture consists of folds, veils, and a few microvilli. Note that one of the cells is standing on its head (b)

a SEM: × *14,500*
b SEM: × *14,200*

Figure 30. Leukemic promyelocyte *(see page 54)*

a This is the typical appearance of a locomotive promyelocyte (pm). The cell body is elongated and shows prominent veils at the front. A constriction at the posterior part suggests the presence of a tail. It is, however, only a temporary feature. The direction of movement is indicated by an arrow (see Figure 23b)

b The shape changes of a locomotive promyelocyte are shown in a set of drawings depicting the same cell with a time interval of 15 s between each drawing

a SEM: × *9700*
b Drawing from a 16-mm film sequence

Figure 31. Leukemic myelocyte *(see page 55)*

a Locomotive activity of this myelocyte is indicated by its round posterior part and the large cytoplasmic veil at the opposite side. At the moment of fixation the large veil was attached to the substrate. The arrow indicates the direction of movement

b The shape changes occurring in a myelocyte during locomotion are illustrated in drawings of the same cell with a 15-s interval between each position

a SEM: × *11,000*
b Drawing from a 16-mm film sequence

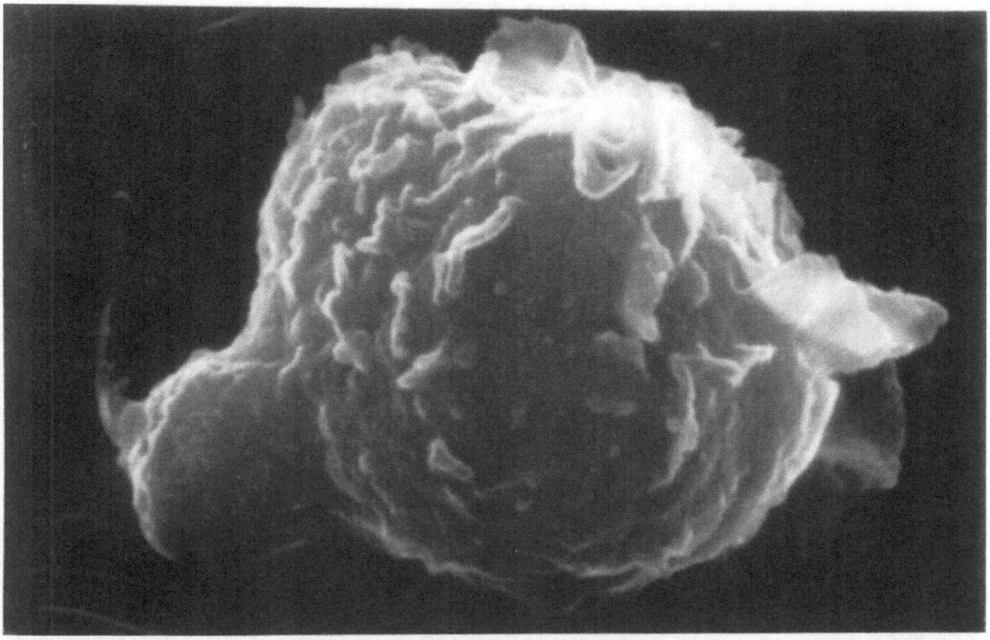

a

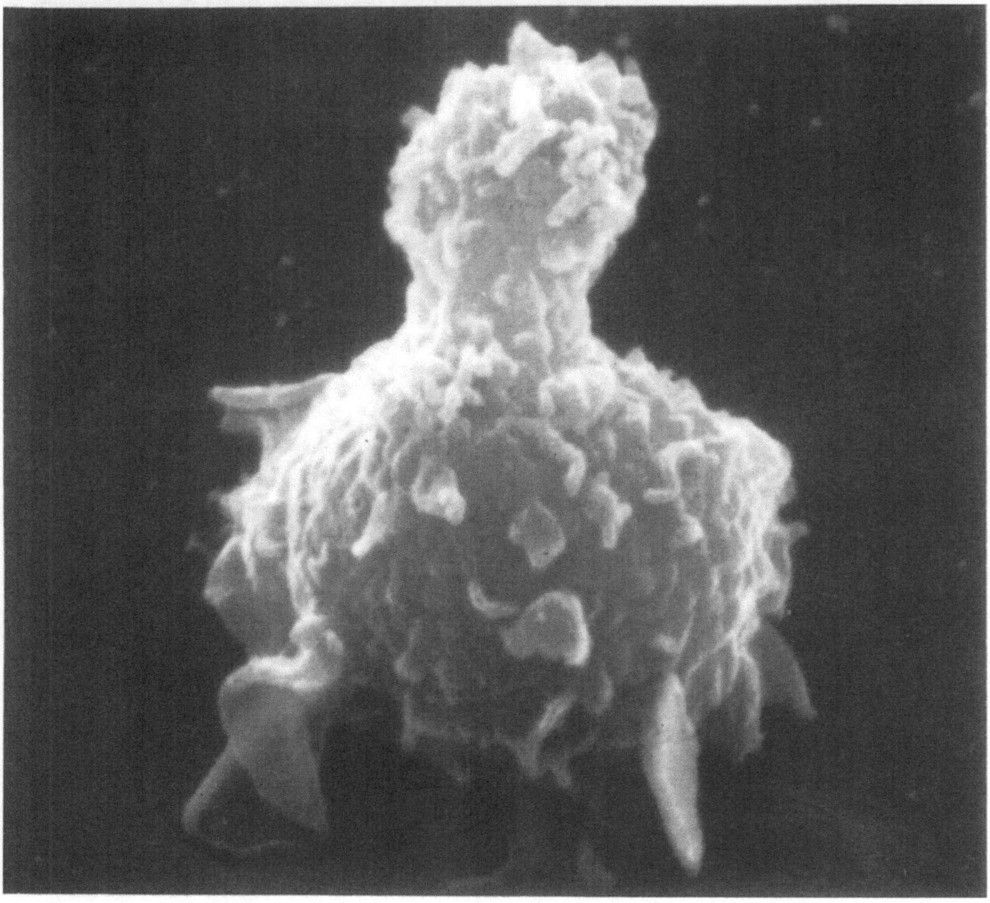

b

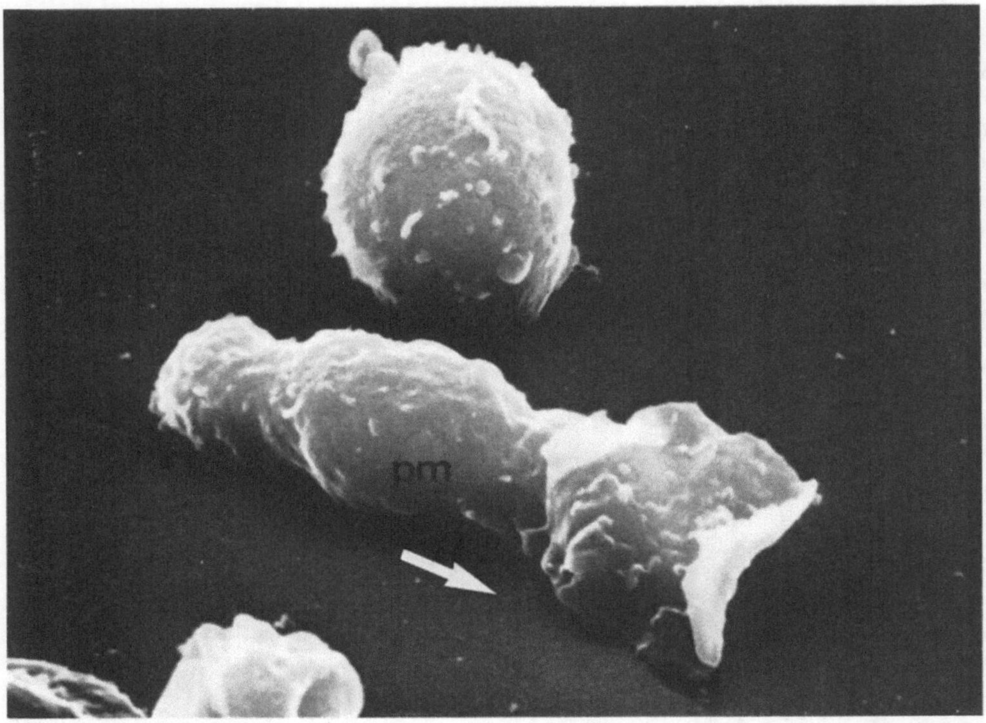

a

SHAPE CHANGE DURING LOCOMOTION

(15 sec interval between two drawings)

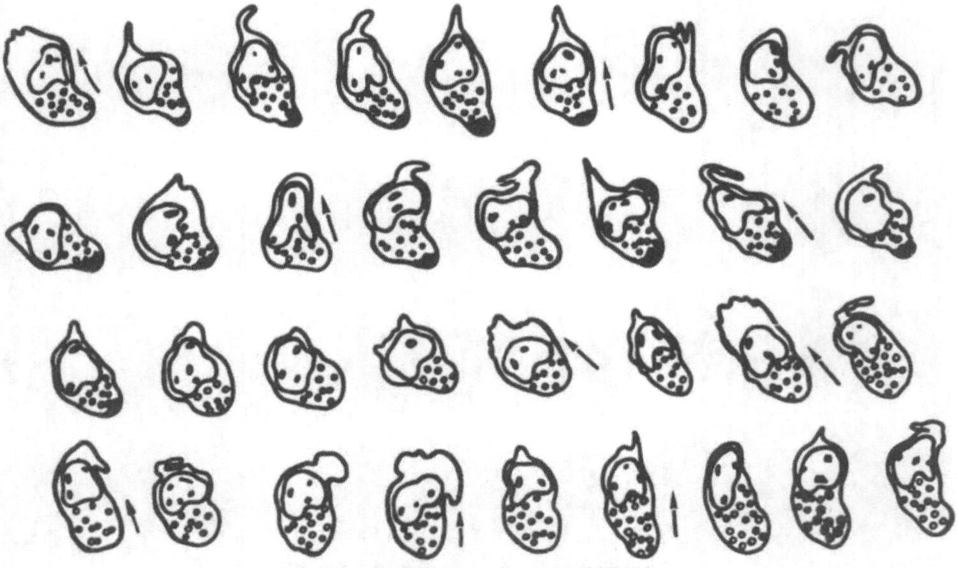

Leukemic promyelocyte (AML)

b

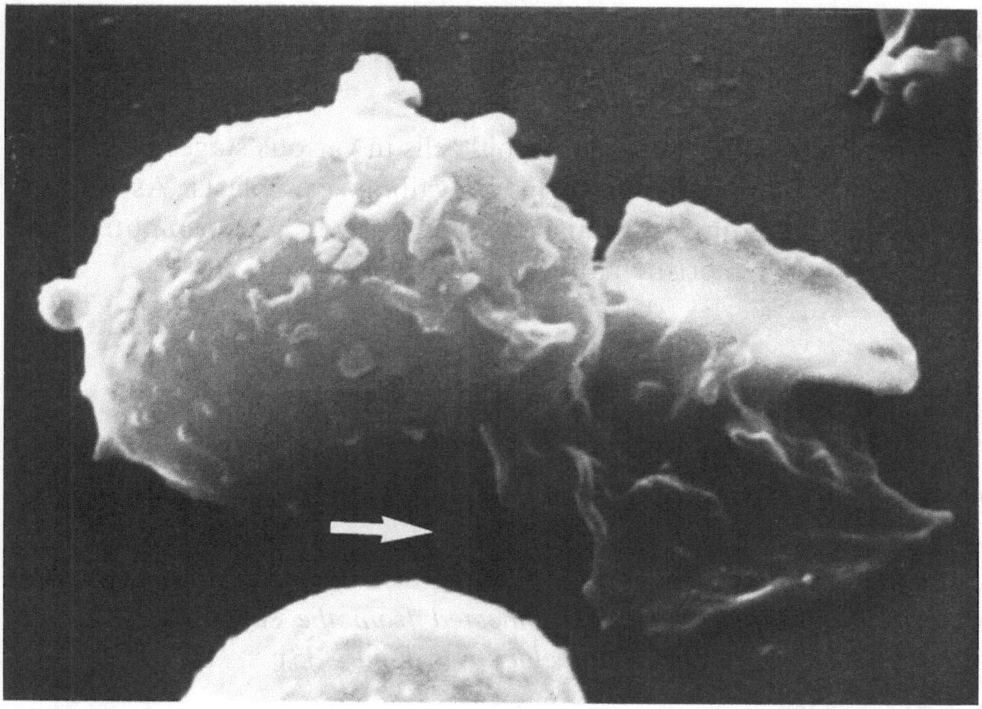

a

SHAPE CHANGE DURING LOCOMOTION

(15 sec interval between two drawings)

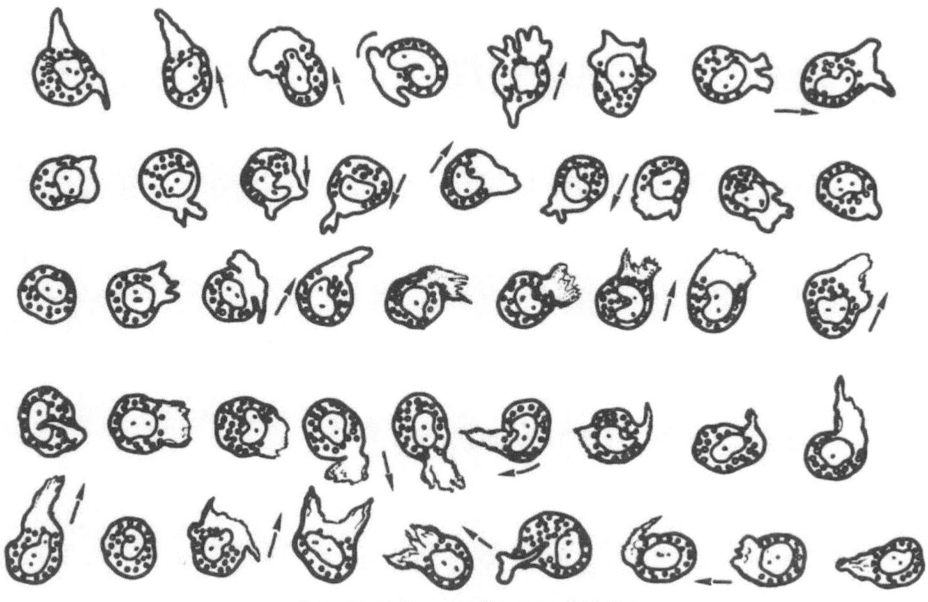

b **Leukemic myelocyte** (AML)

This is another leukemia with myeloid cells in various stages of maturation, the more differentiated forms prevailing over blasts. As was the case for the preceding leukemia, most of the cells are locomotive and display their characteristic pattern of movement.

Figure 32. Leukemic promyelocyte and myelocyte

These are the two main representatives of this leukemia: promyelocyte (a) and myelocyte (b). Both cells are shown in a position suggesting locomotion. The promyelocyte was fixed in a momentary configuration when the anterior part was disconnected from the substrate. A similar situation is shown for the promyelocyte in Figure 33c

a SEM: ×8400
b SEM: ×9800

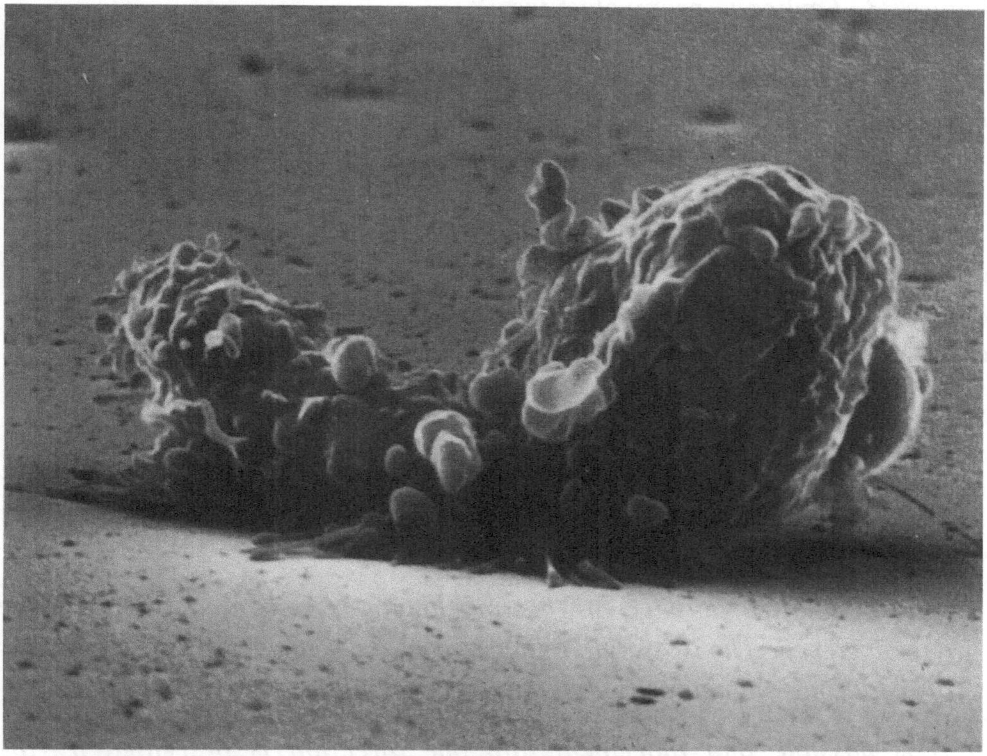
a
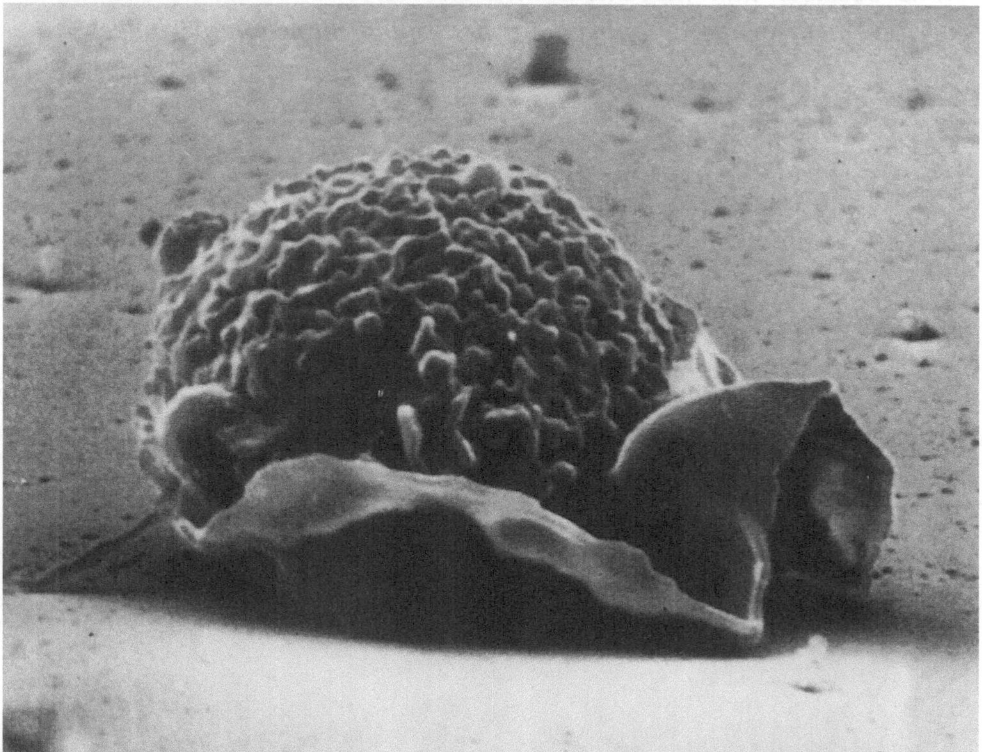
b

Figure 33. Leukemic promyelocytes

a and c Locomotive promyelocytes seen with a tilt angle of 85°

b Drawing of the pathway of a locomotive promyelocyte that covered the track of 115 μm within 7.2 min. The velocity of the cell was in the order of 16 μm/min

a SEM: × 7200
b Drawing from a 16-mm film sequence
c SEM: × 7300

Figure 34. Leukemic myelocytes *(see page 60)*

a The locomotive activity of this myelocyte is indicated by the large one-sided veil. The two small extensions at the rear end are not remnants of a tail but momentarily existing cytoplasmic folds

b It is possible that movement into one direction is prevented by the arrangement of the cytoplasmic veils around the whole periphery

a SEM: × 9800
b SEM: × 8100

Figure 35. Leukemic myelocyte *(see page 61)*

Another representation of a locomotive myelocyte in its characteristic configuration. The arrow points into the direction of movement

SEM: × 12,700

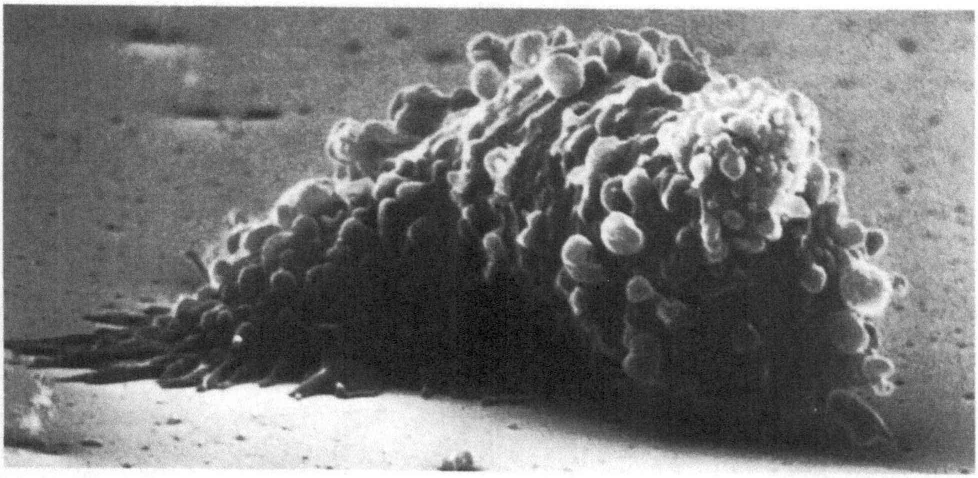

a

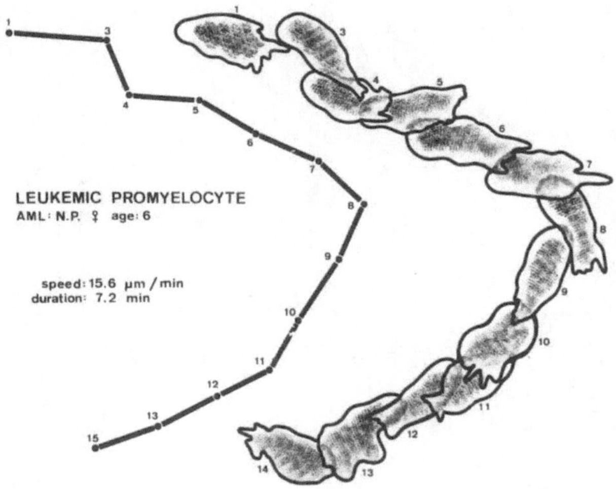
LEUKEMIC PROMYELOCYTE
AML: N.P. ♀ age: 6
speed: 15.6 μm / min
duration: 7.2 min

b

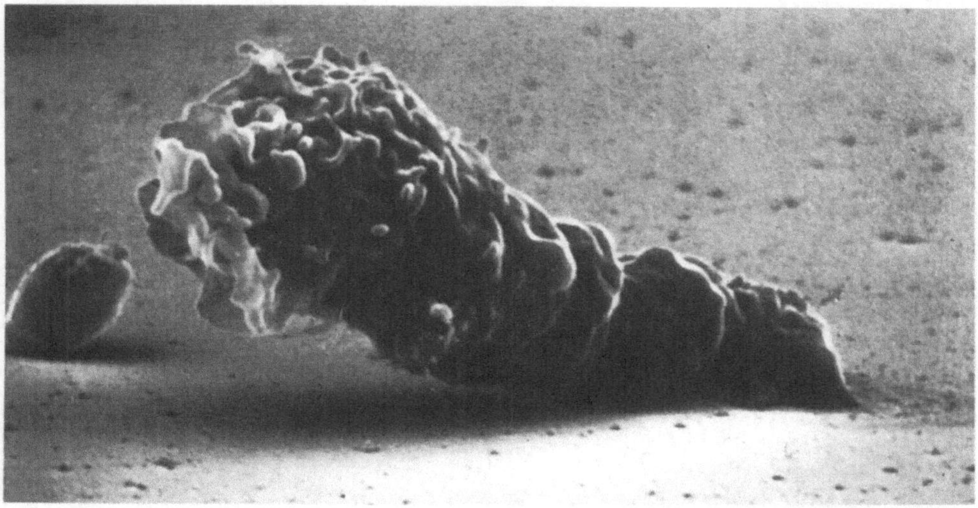

c

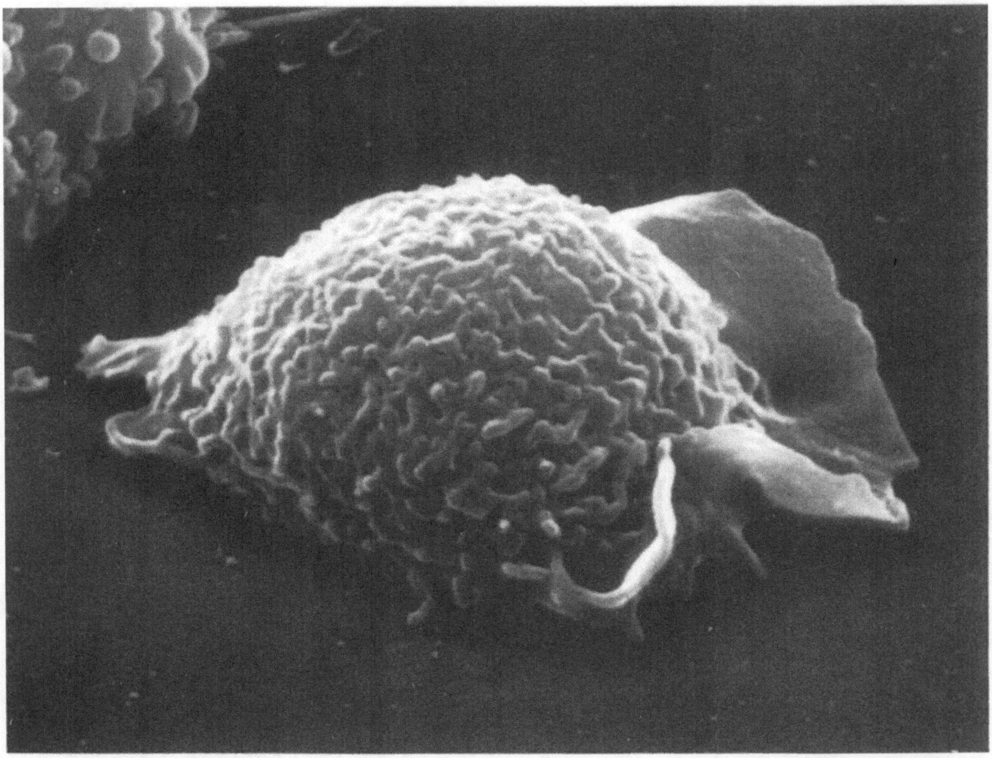
a

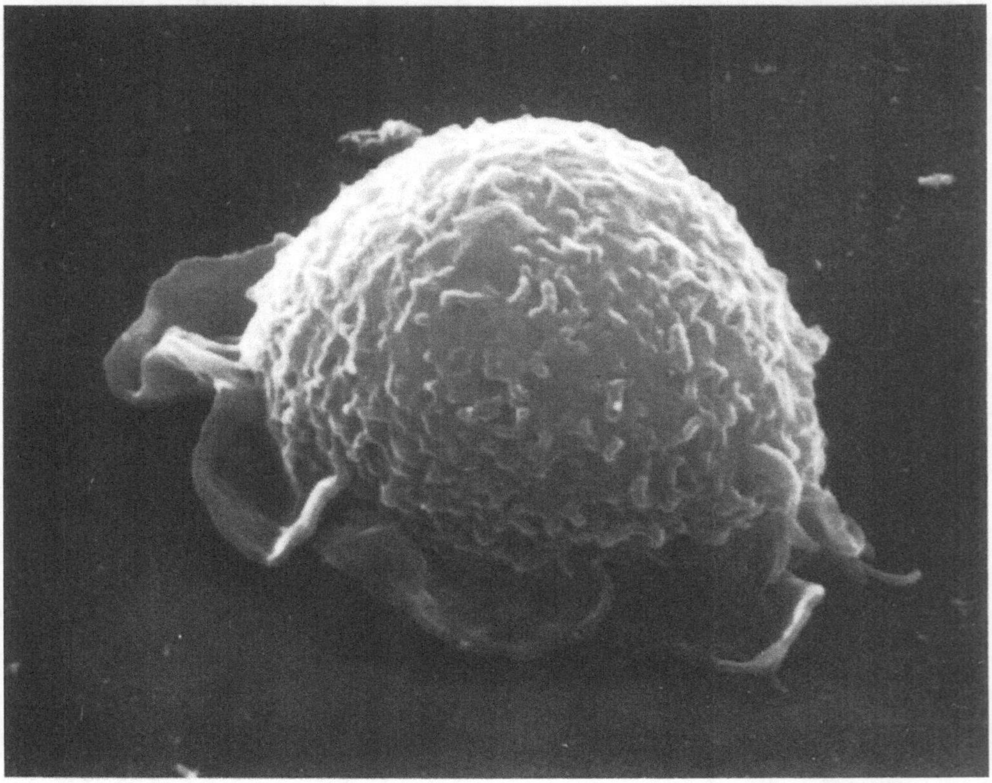
b

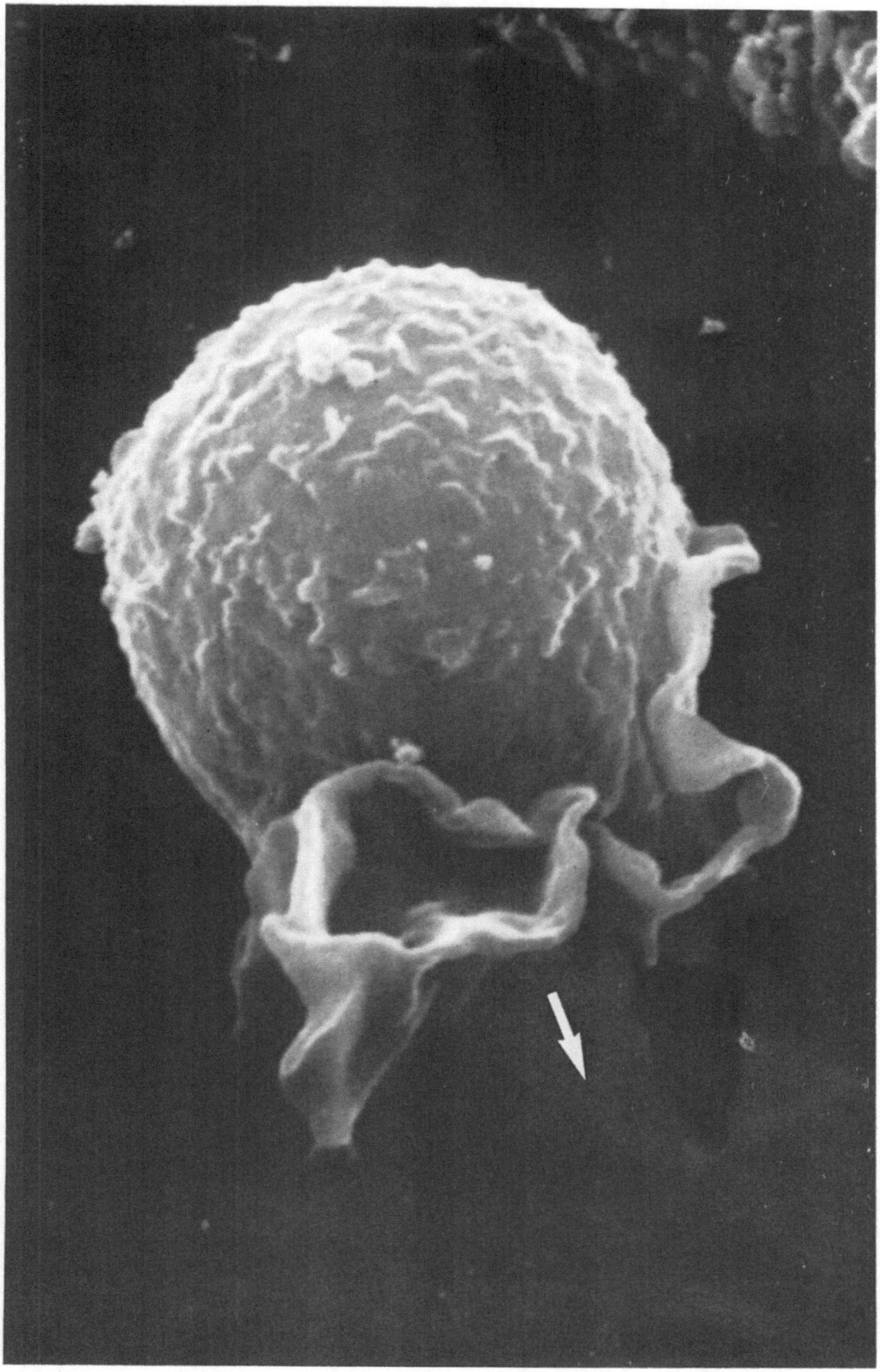

2. Chronic Myeloid Leukemia (CML)

This chronic myeloid leukemia is characterized by the scarcity of myeloblasts, and the presence of cells in various degrees of differentiation with promyelocytes and myelocytes dominating the population. On-spot motility and locomotion of the different cell types result in a great variability in shape and behavior, reflected in SEM photographs and verified by time-lapse films.

Figure 36. Leukemic population

Different representatives of the myeloid series engaged in various on-spot and locomotive activities are recognizable. Besides a spheric cell that could be a myeloblast (mb?), there are a promyelocyte (pm), myelocytes (my), and thrombocytes (th). It is typical for the myelocytes in this leukemic population that during locomotion they can flatten to a considerable degree. One is indicated by an arrow

a SEM: ×3100
b SEM: ×3400

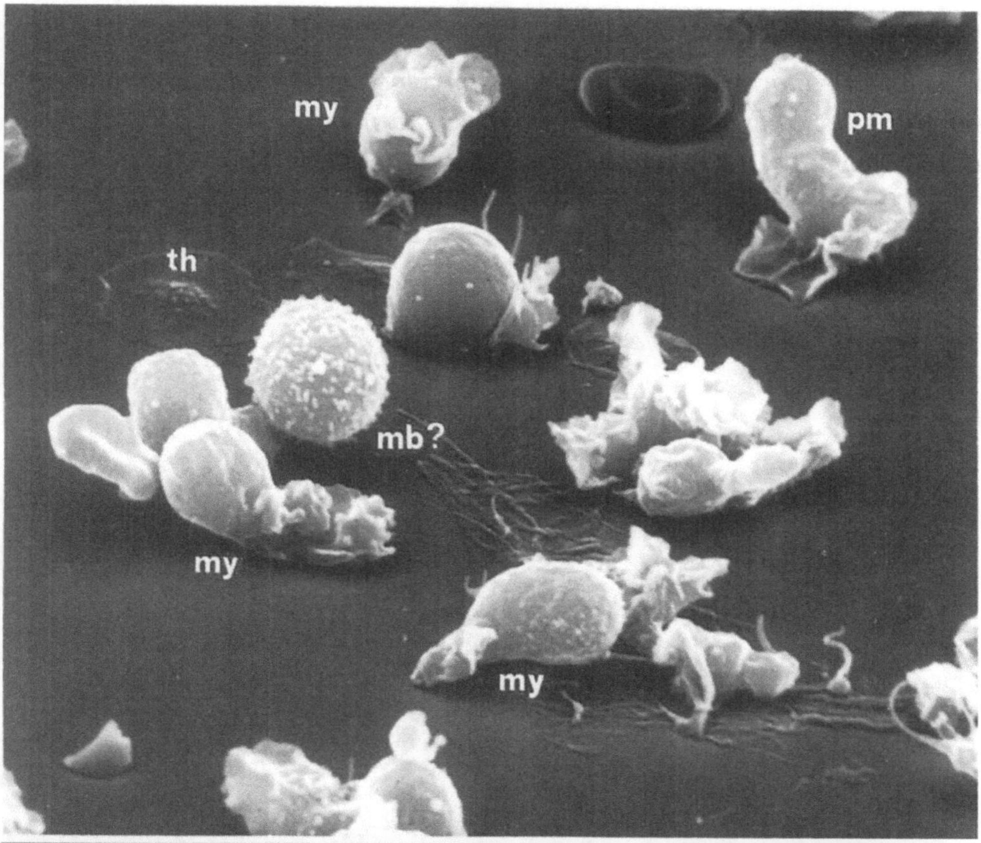

a

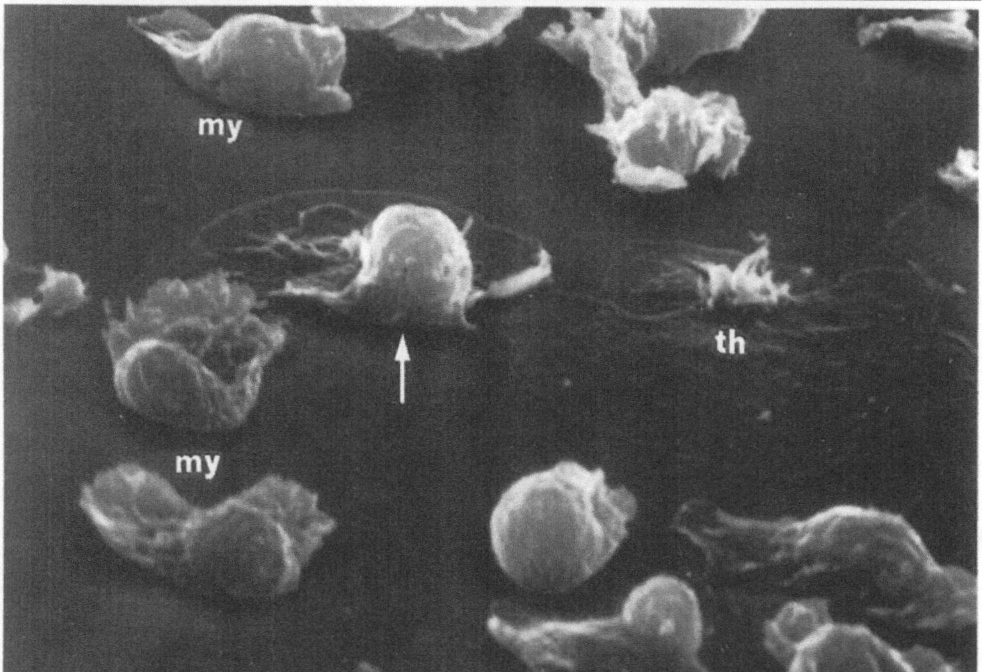

b

Figure 37. Leukemic promyelocytes

The elongated form of a locomotive promyelocyte with a large cytoplasmic extension pointing into the direction of movement is registered by time-lapse cinematography (a) and presented in detail by SEM (b)

a Phase contrast: × 2700
b SEM: × 10,600

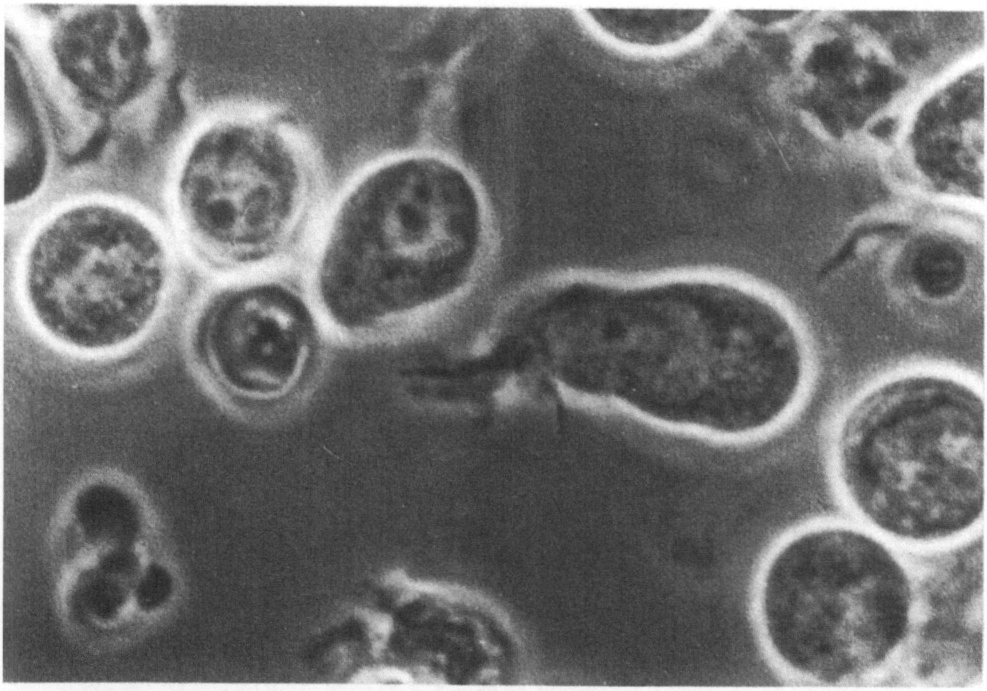
a

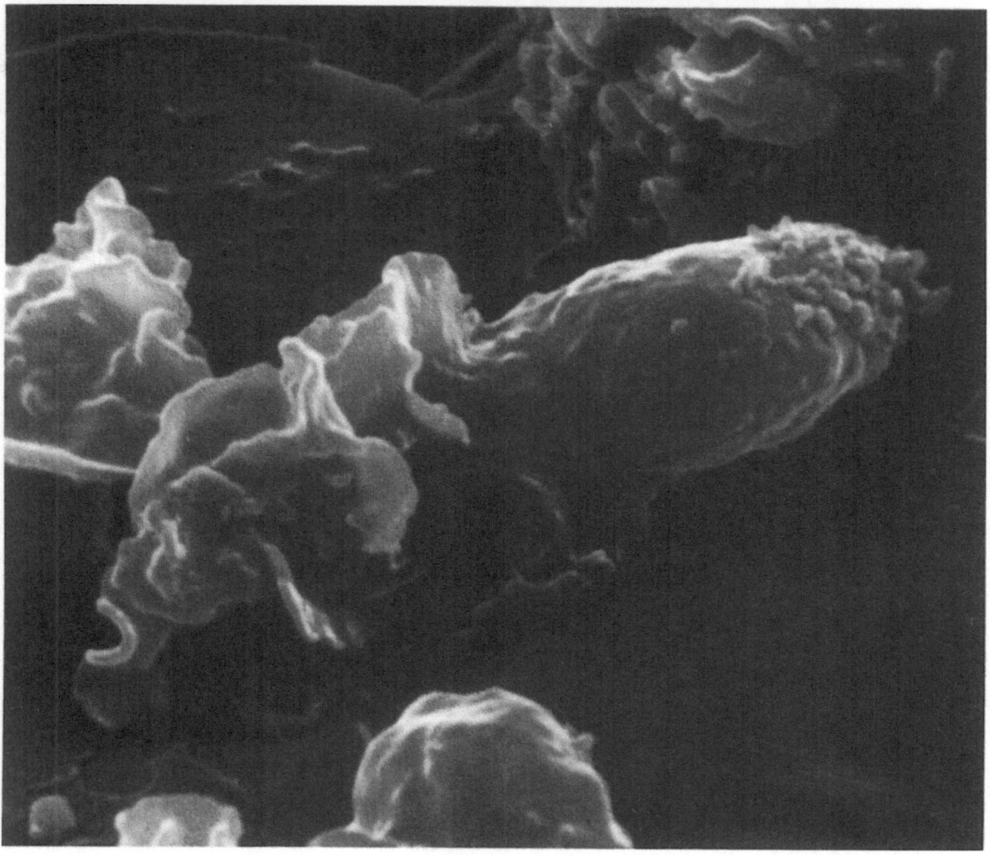
b

Figure 38. Leukemic promyelocyte and myelocyte

a This is a promyelocyte in its characteristic configuration assumed during locomotive activity. The direction of movement is indicated by the large anterior cytoplasmic extensions that have established contact with the substrate while the body, at the moment of fixation, was lifted into the medium

b As the veil-like extensions do not point into one direction, it can be assumed that this myelocyte performed, at the time of fixation, the special surface motility typical for this cell type

a SEM: × 14,100
b SEM: × 14,100

Figure 39. Leukemic myelocytes *(see page 70)*

The possibility of a dynamic morphologic interpretation, based on the combination of MCM and SEM, is well documented in the case of these locomotive myelocytes (my) with their characteristic round posterior part and their broad cytoplasmic fringes in front. The direction of movement is indicated by an arrow for one of the cells

a Phase contrast: × 2100
b SEM: × 14,100

Figures 40, 41, 42. Leukemic myelocytes *(see pages 71, 72, 73)*

Different momentary positions of locomotive myelocytes demonstrate the great versatility in shape and surface architecture of this cell type

40 SEM: × 21,100
41a SEM: × 15,400
41b SEM: × 7400
42a SEM: × 13,600
42b SEM: × 14,500

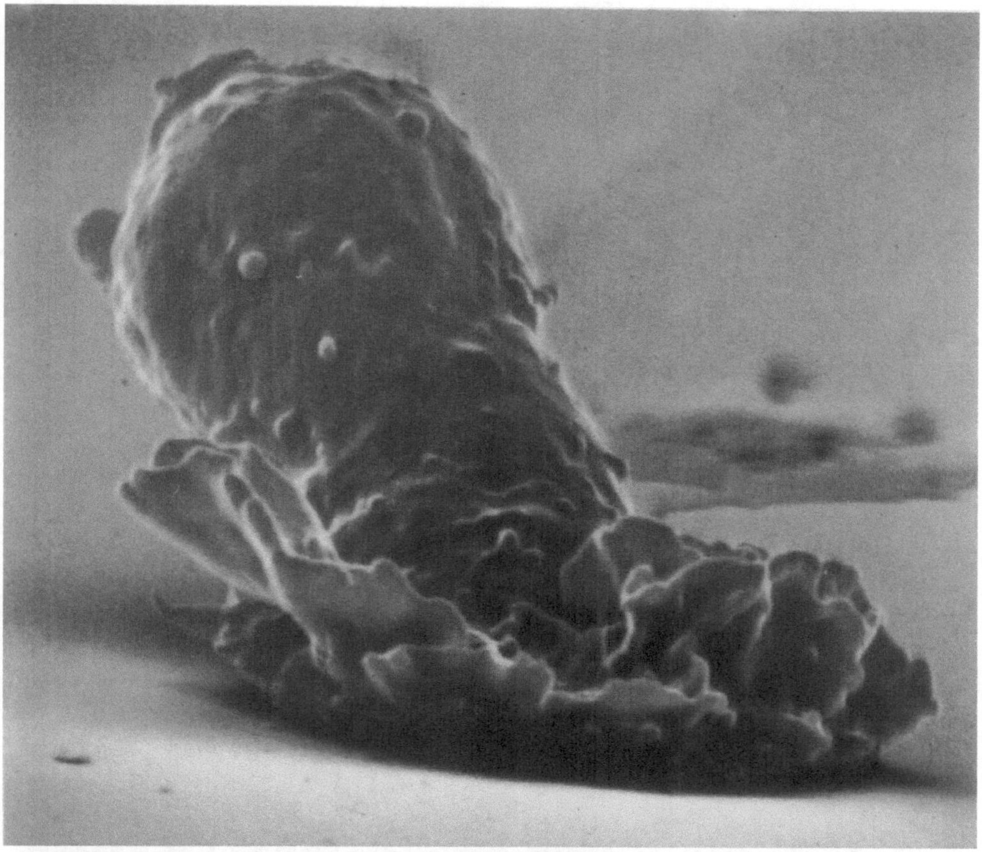

a

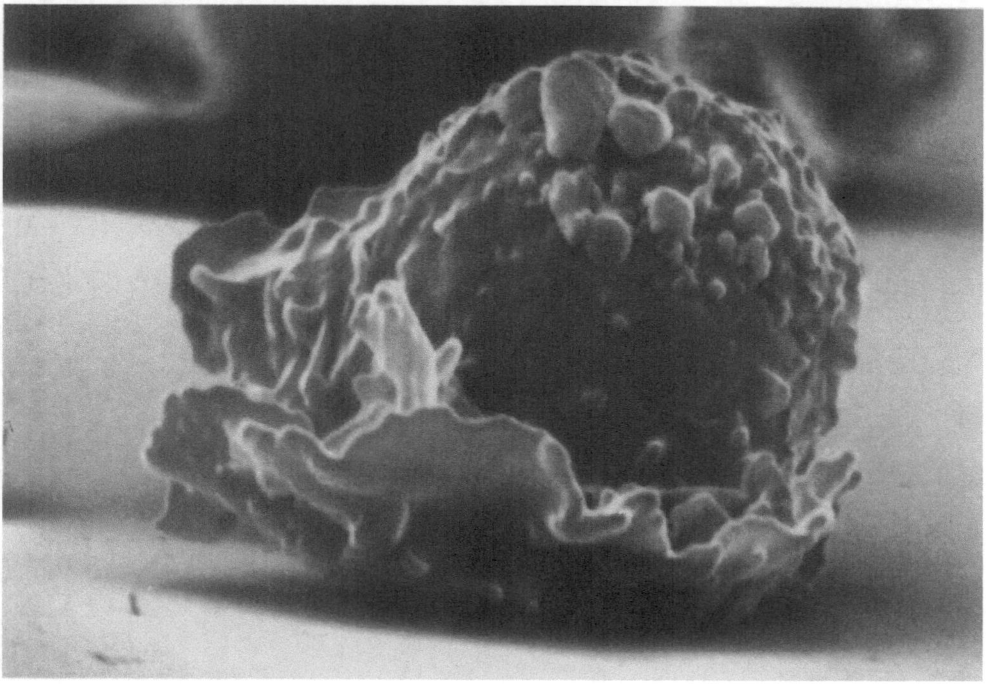

b

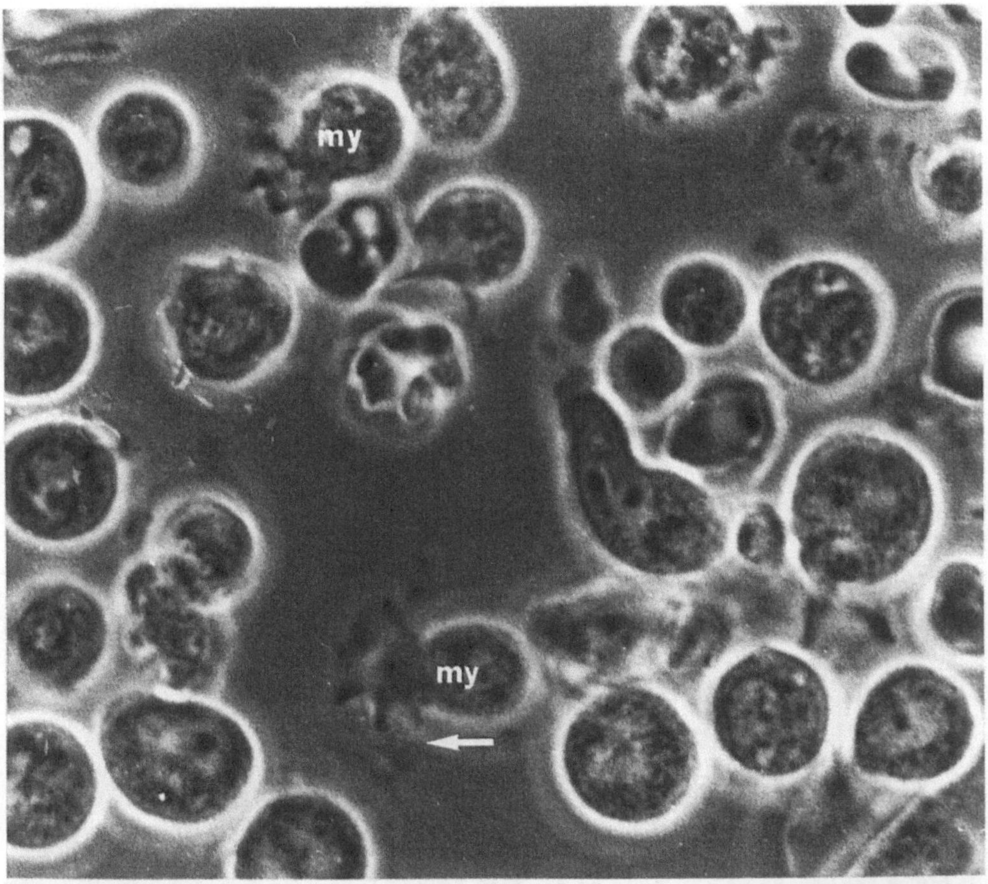

a

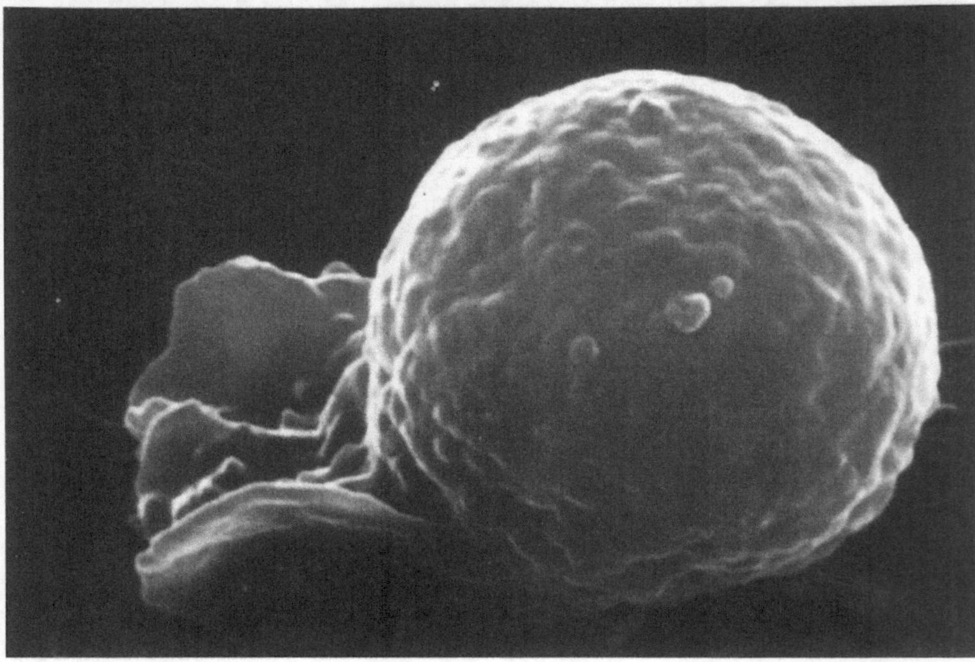

b

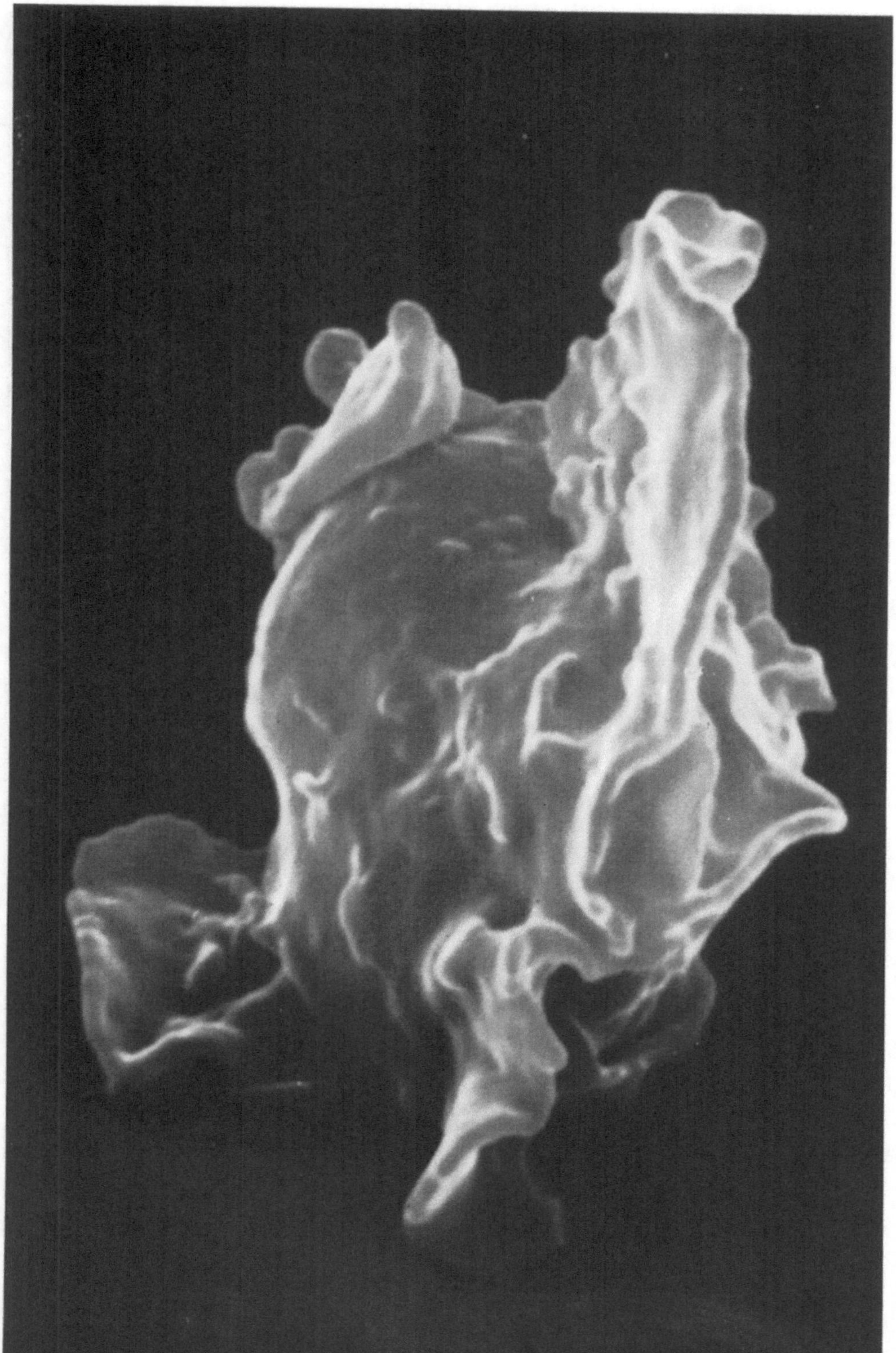

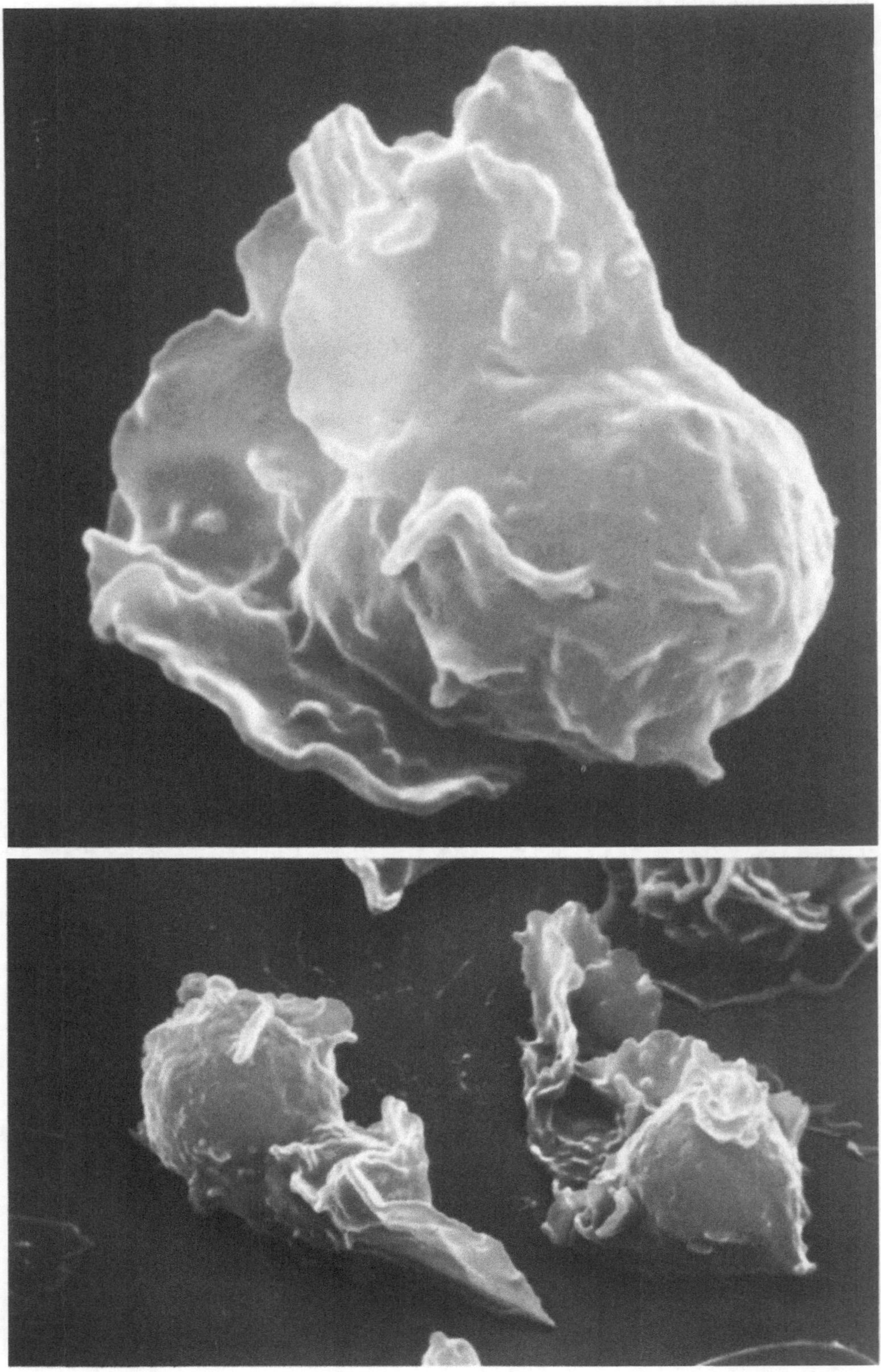

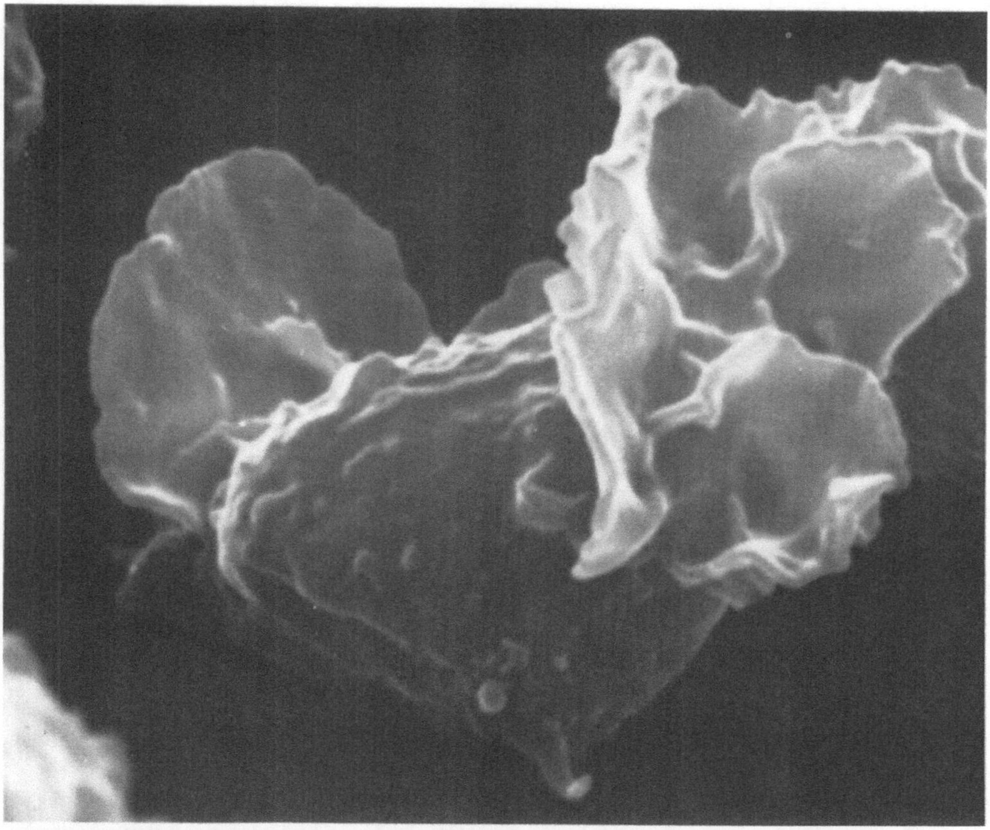

a

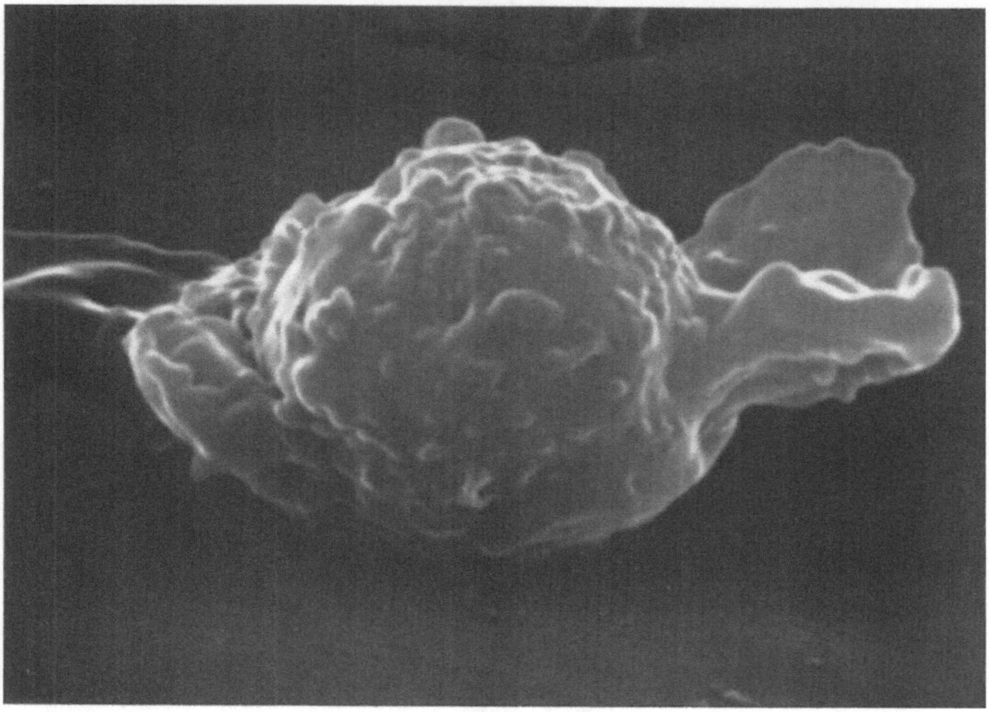

b

3. *Acute Myelomonocytic Leukemias (AMML)*

In this series of myeloid leukemias, monocytes represent a considerable part of the population. In case 1, myeloblasts and monocytes, both locomotive, are the main representatives.

Figure 43. Leukemic myeloblast and leukemic monocyte

The two main cell types are shown: a myeloblast (a) in its blast configuration and a monocyte (b). The differences of the two cells, both in shape and surface extensions, are well documented by SEM at a tilt angle of 85°. Note the especially large veils of the advancing monocyte (b)

a SEM: × 8000
b SEM: × 8900

Figure 44. Leukemic myeloblast *(see page 78)*

The polarized shape of blast cells, characteristic for locomotion in time-lapse films, is also preserved in this TEM preparation. Note the similarity of the tail to that in Figure 45b

TEM: × 14,800

Figure 45. Leukemic myeloblasts *(see page 79)*

The cells exhibit the polarized form of locomotive blast cells. Arrows indicate the direction of movement

a SEM: × 15,700
b SEM: × 9800

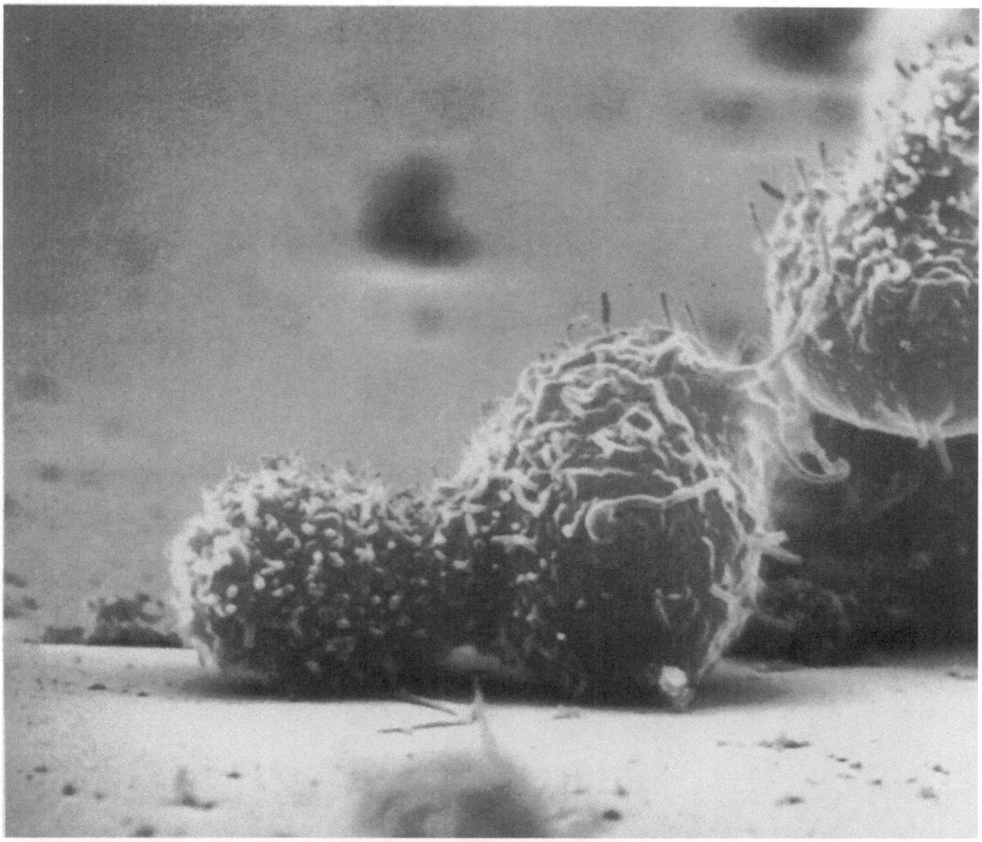

a

b

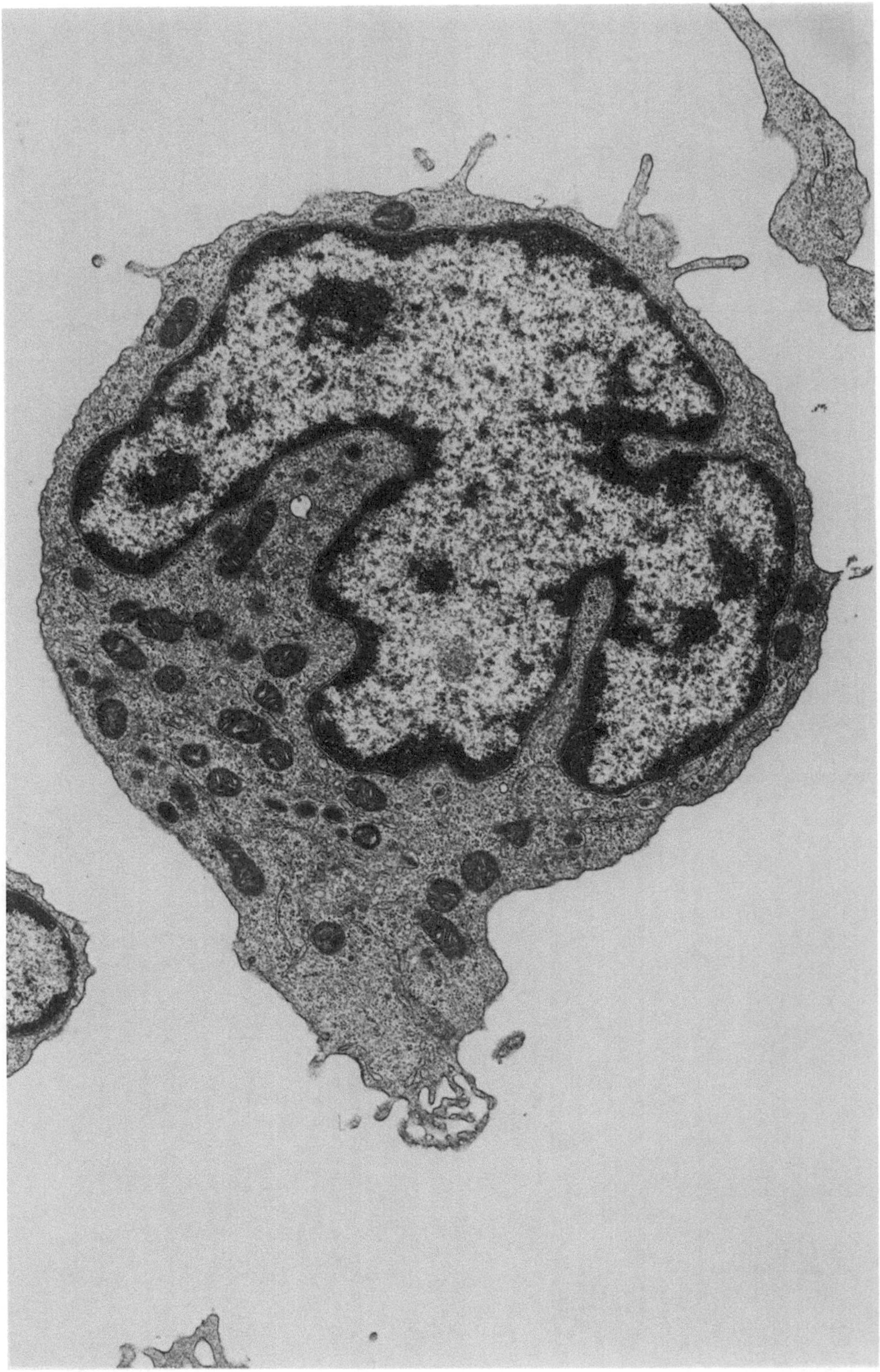

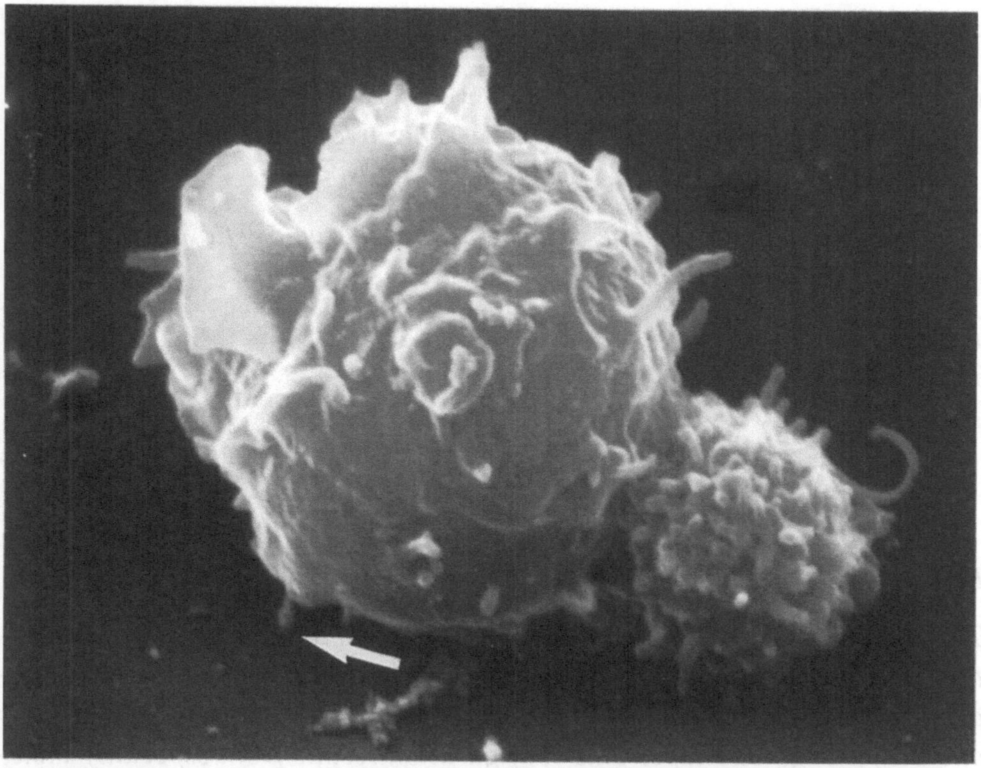

a

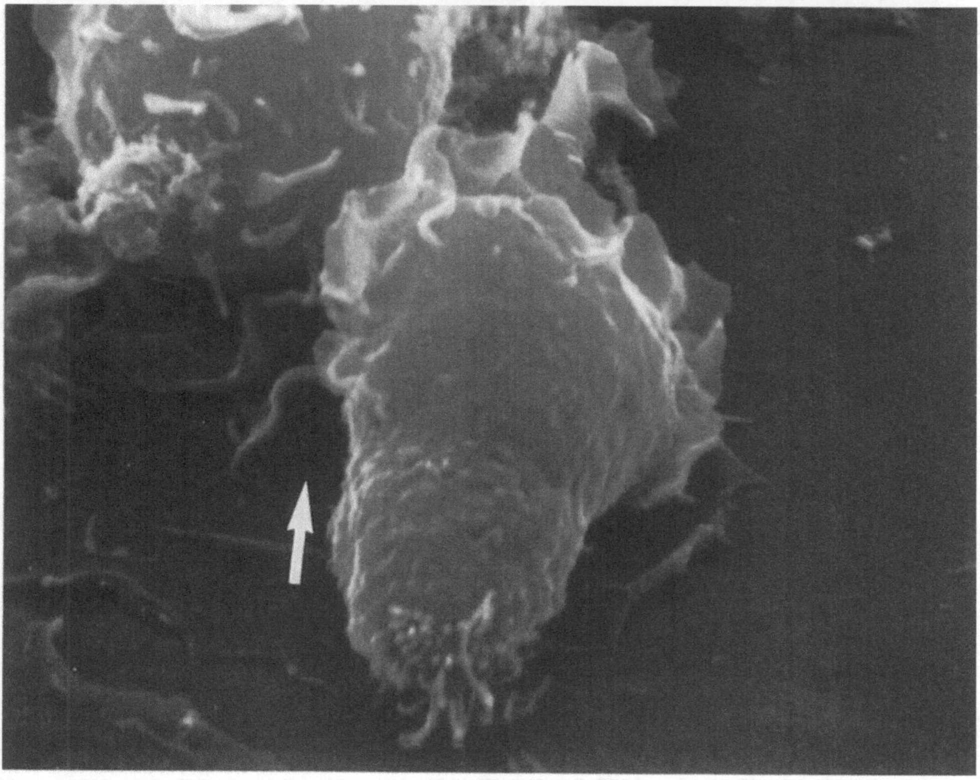

b

Figure 46. Leukemic monocytes

a The SE micrograph shows several well-spread monocytes with small-sized ruffles around their periphery. The arrow indicates a monocyte that is still spheric with its typical large folds

b Interdigitating cytoplasmic extensions are characteristic for the monocytes of this population

a SEM: × 3500
b TEM: × 3700

Figures 47, 48. Leukemic monocytes *(see pages 82, 83)*

This set of photographs again demonstrates the similarity of results obtained with cinematography and electron microscopy. Among spread monocytes with their typical, centrally positioned folds there are also cells of a very elongated shape. This configuration results from especially adhesive properties of the tail during locomotion as documented by time-lapse film sequences

47a SEM: × 3200
47b Phase contrast: × 2400
48a SEM: × 7300
48b Phase contrast: × 3300

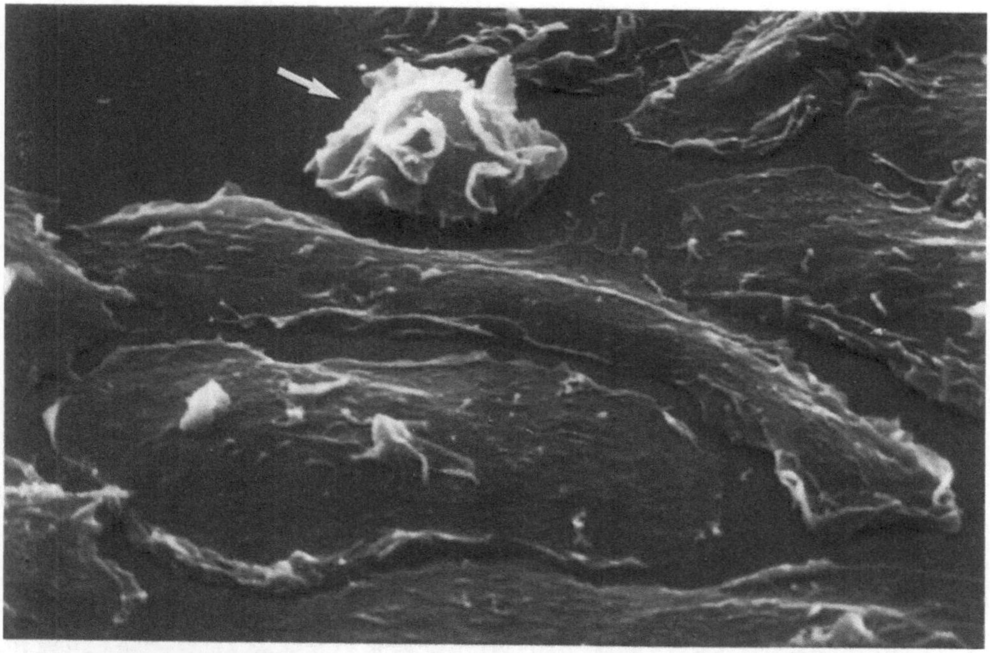
a

b

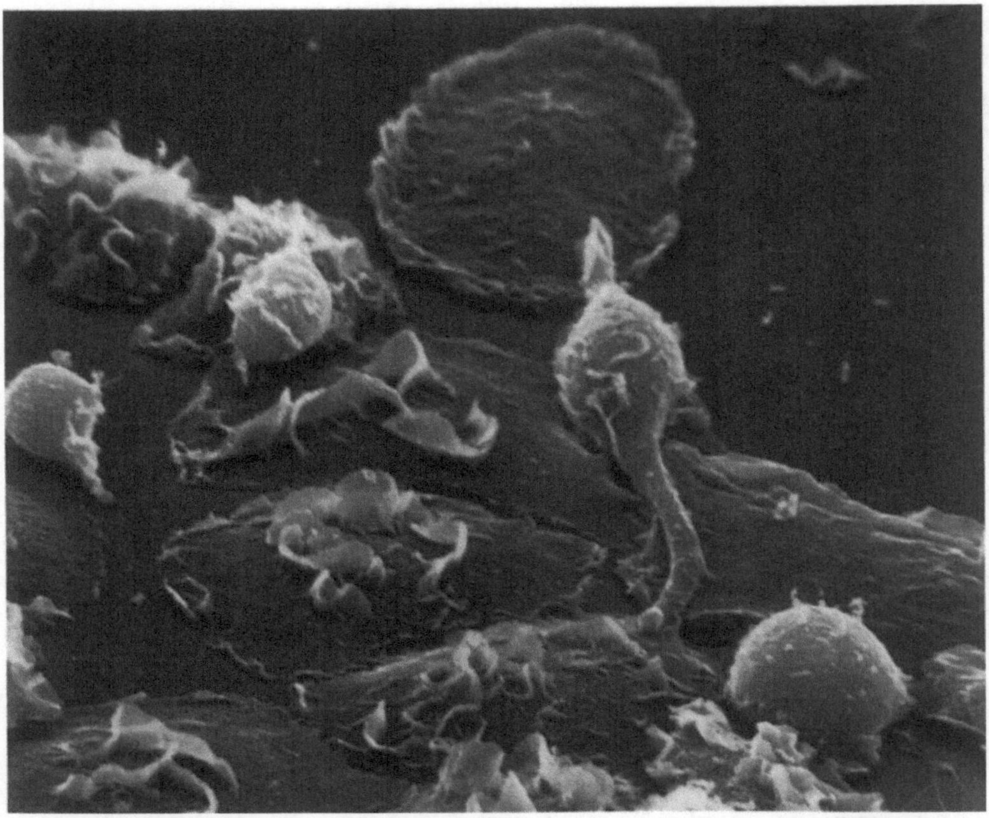

a

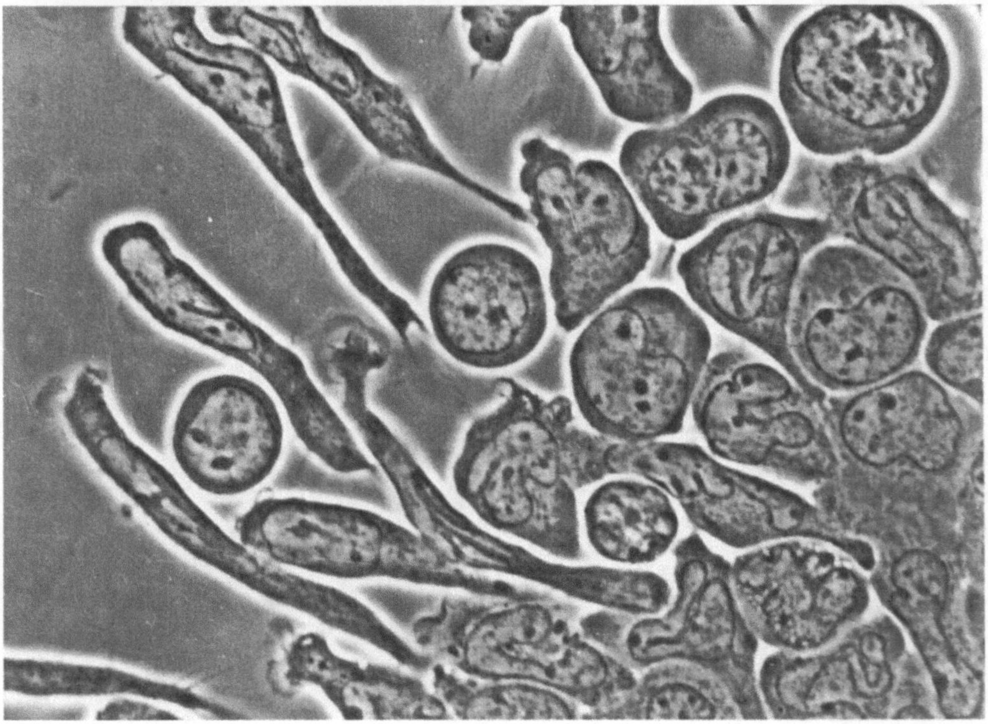

b

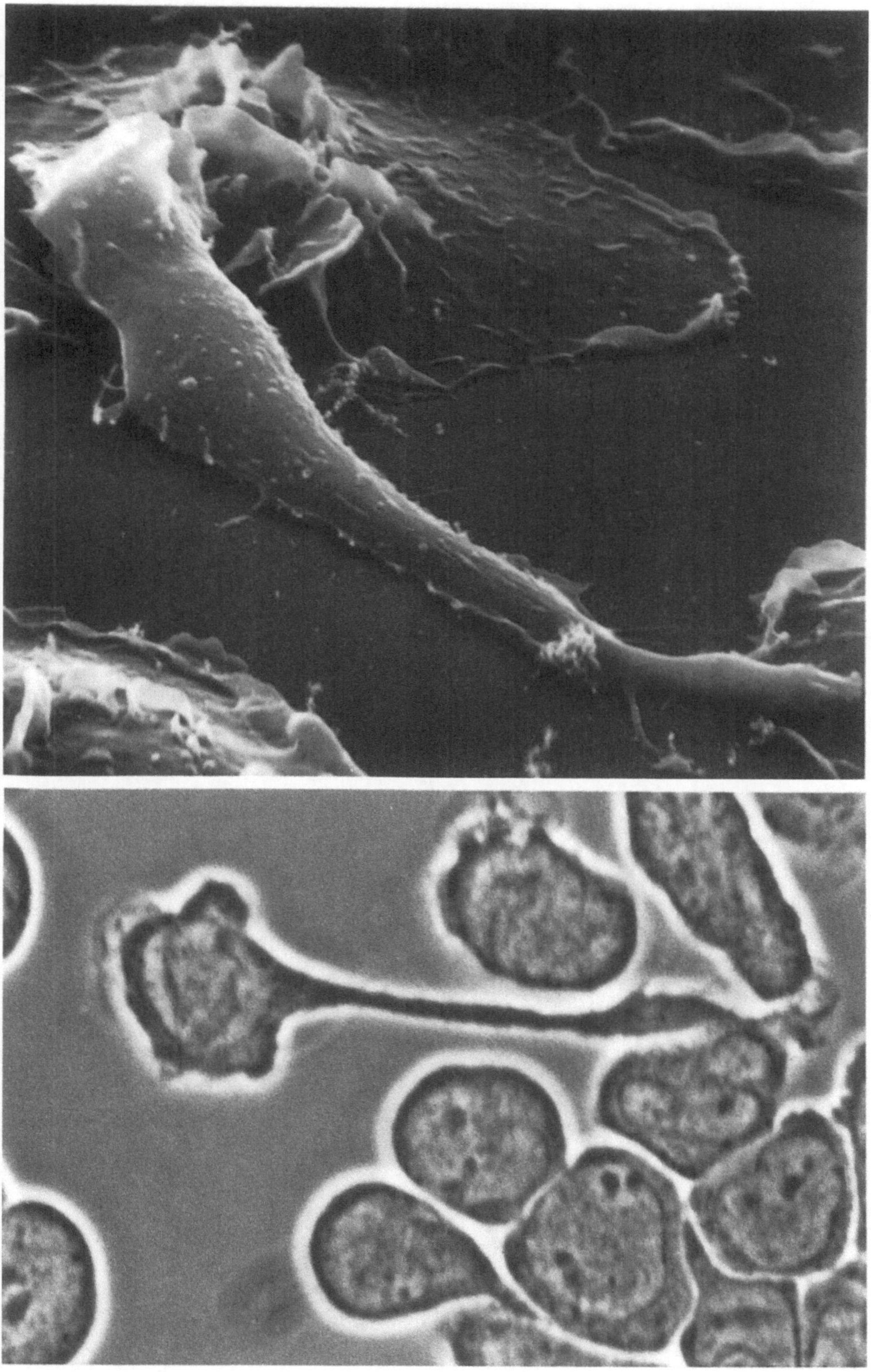

a

b

Figure 49. Leukemic monocytes

Two further examples of locomotive monocytes with long tails and impressive folds covering the cell body. The special adhesiveness of the tail is suggested by its spread form on one of the cells (b)

a SEM: ×8200
b SEM: ×9200

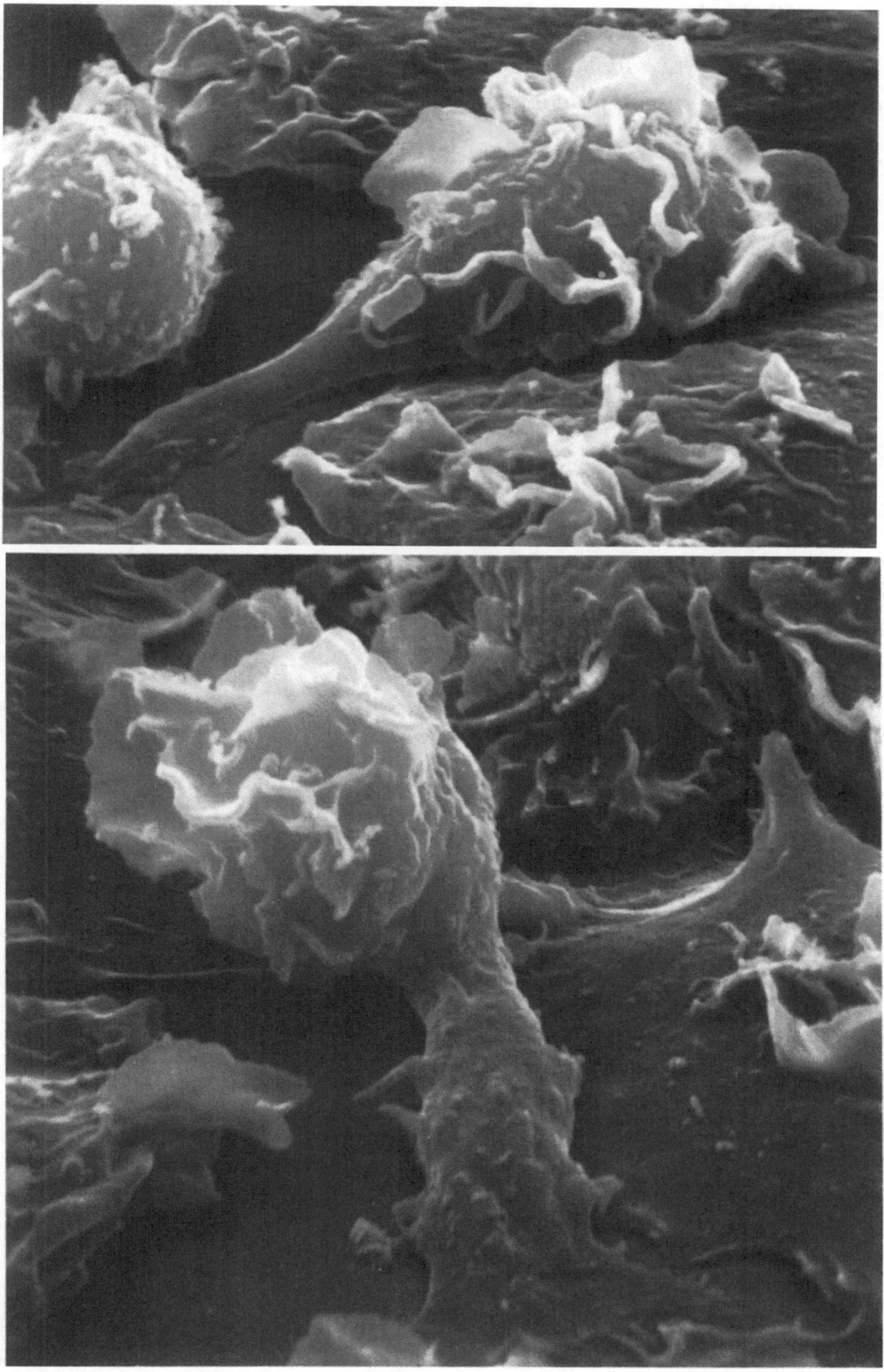

a

b

Figure 50. Leukemic monocytes

Arrows indicate two monocytes in a position compatible with locomotion on top of other spread monocytes

a SEM: × 4600
b SEM: × 4500

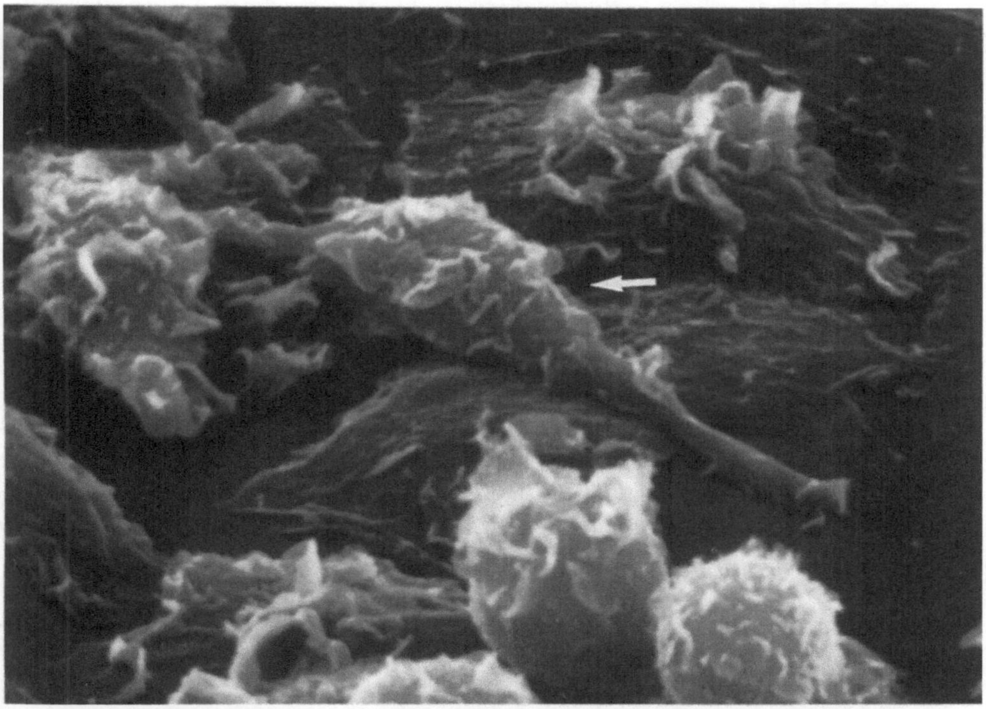

a

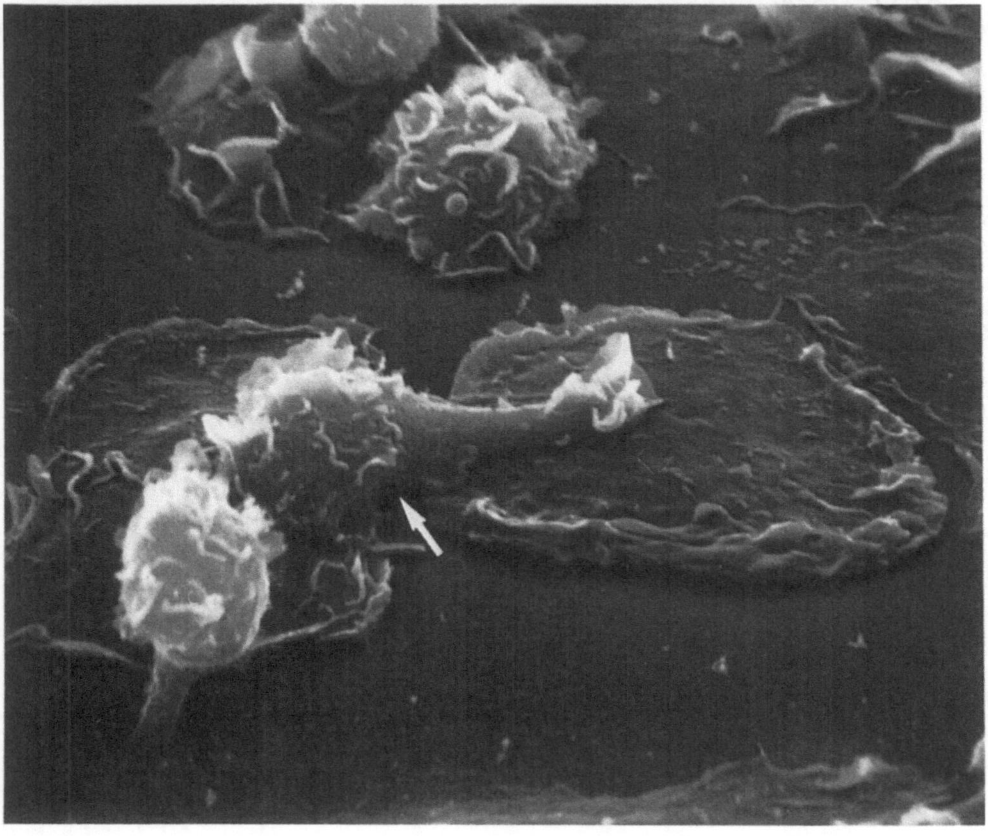

b

Besides myeloblasts and monocytes, this leukemia also contains promyelocytes and myelocytes. All cells are partly engaged in surface motility or in locomotion.

Figure 51. Leukemic population

a Some of the cellular representatives can be seen: monocytes (mo), larger and smaller spheric cells that cannot be identified by SEM, also many thrombocytes (th), spread and nonspread

b This monocyte has asymmetrically arranged folds and conspicuous attachment filaments at the tail-like segment

a SEM: × 3500
b SEM: × 7200

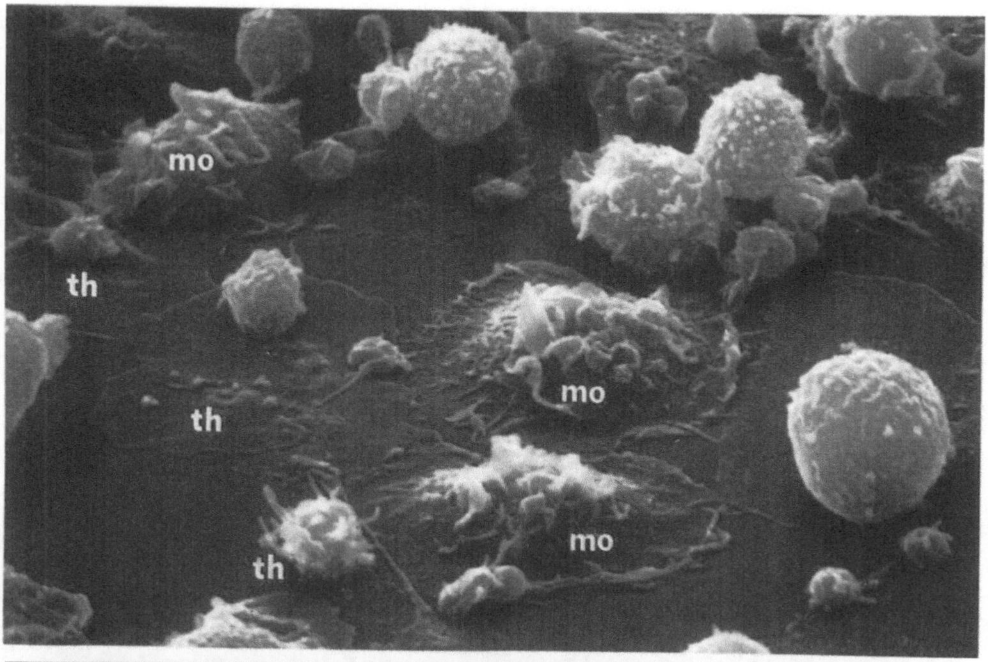

a

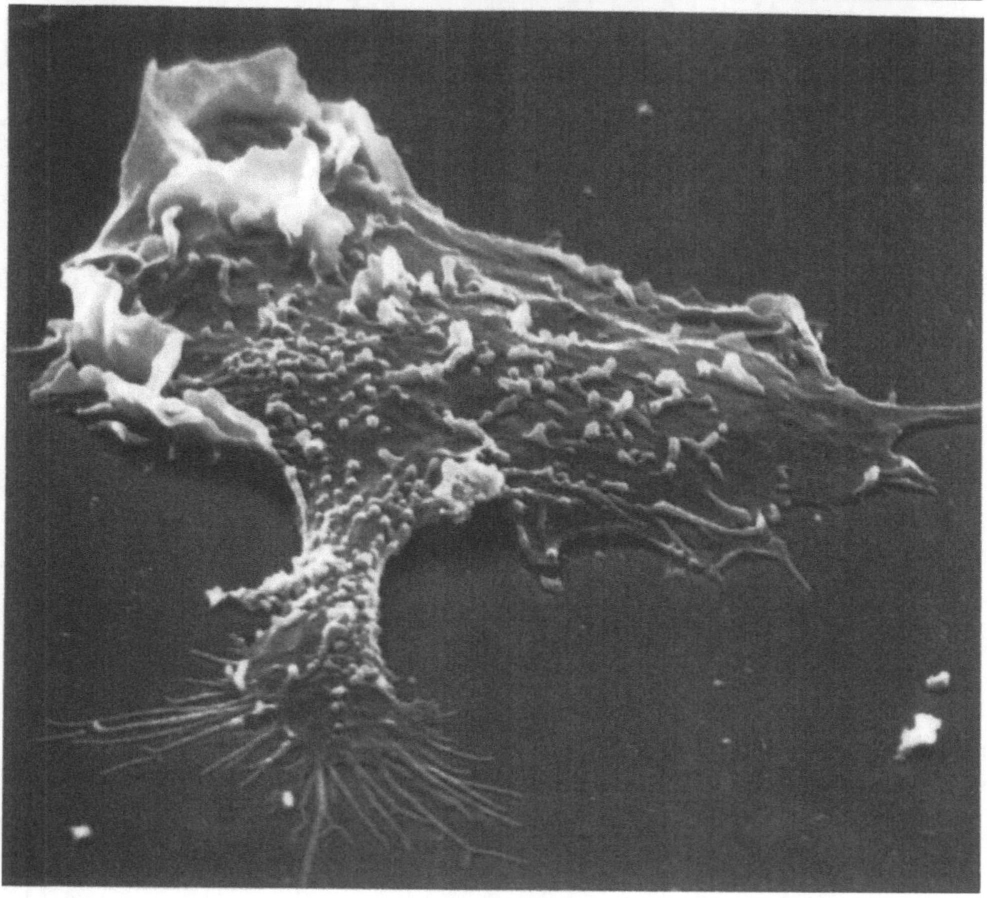
b

Figure 52. Leukemic cells

Two configurational aspects are presented, round (a) and polarized (b). While the spheric cells (a) cannot be identified with certainty, the polarized configuration is suggestive of the blast nature of the cell

a SEM: × 9000
b SEM: × 16,600

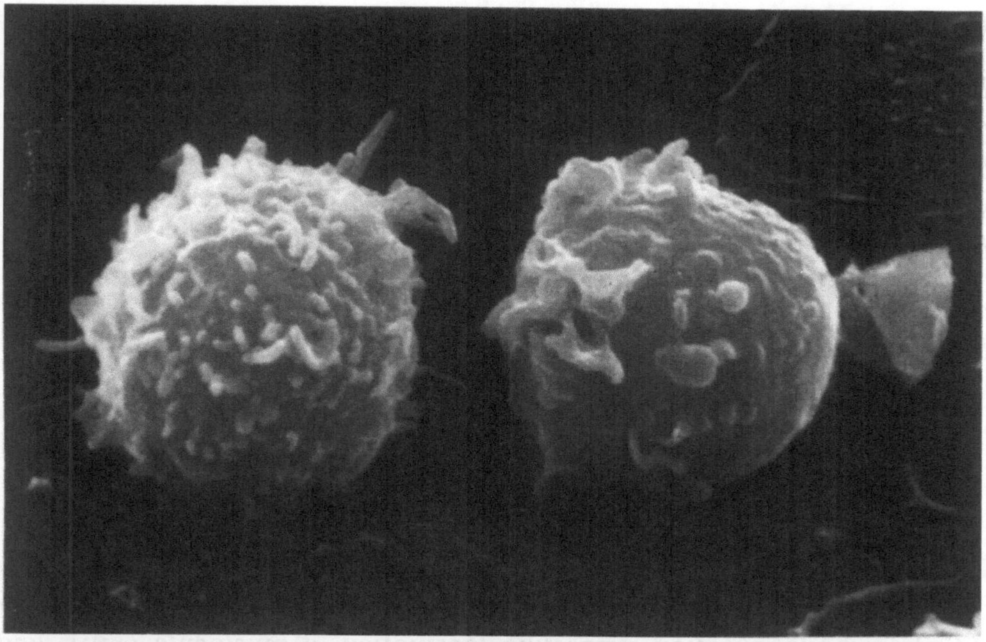
a

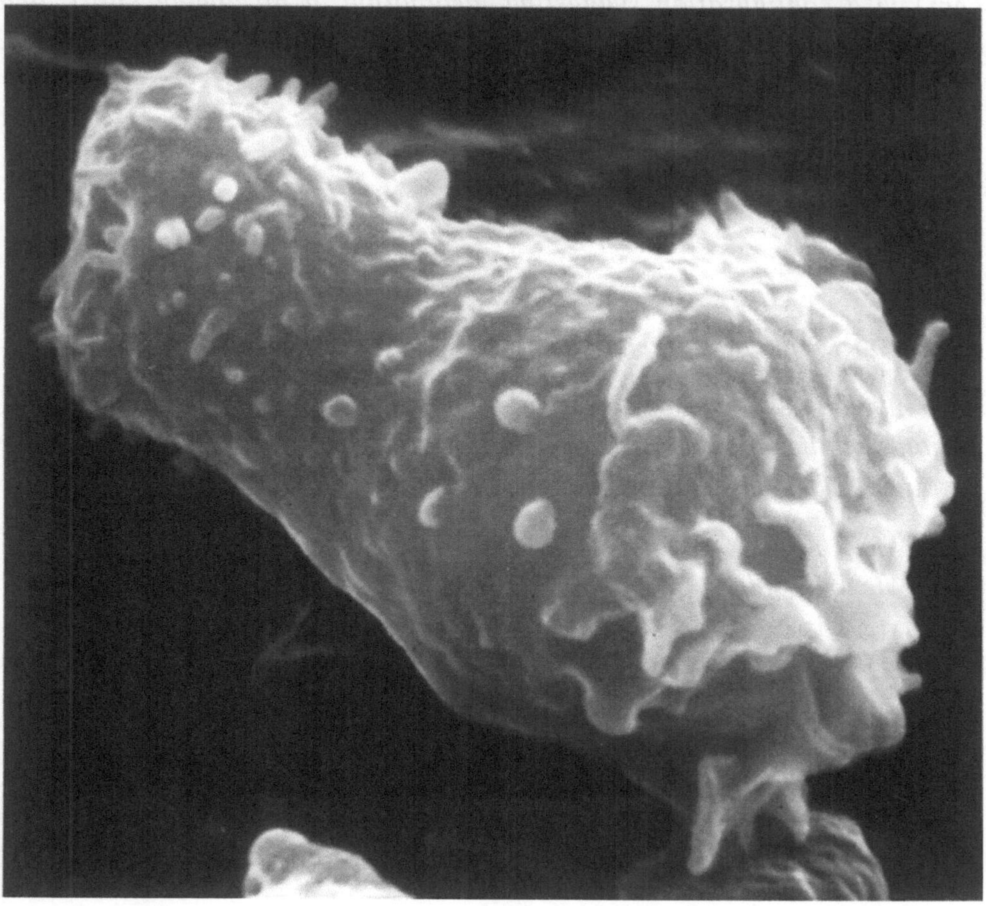
b

Figure 53. Leukemic promyelocytes

In the Nomarski photograph from a time-lapse sequence (a), the locomotive promyelocyte (pm) can be distinguished from the other cell types by its size and elongated form. The veils at the front are especially well seen in the SE micrograph (b). The direction of movement is indicated by arrows

a Nomarski contrast: × *1400*
b SEM: × *10,300*

Figure 54. Leukemic population *(see page 94)*

A higher magnification of part of the population shows a promyelocyte (pm) in a configuration suggesting locomotion, a myelocyte (my), and a thrombocyte (th) located on the surface of a spread monocyte (smo). Part of a less spread monocyte (mo) is also visible

SEM: × *9400*

Figure 55. Leukemic myelocytes *(see page 95)*

Two myelocytes adhering to spread monocytes show the polarized configuration of either locomotion or on-spot motility. In this case, SEM cannot distinguish between the two activities

a SEM: × *14,600*
b SEM: × *15,500*

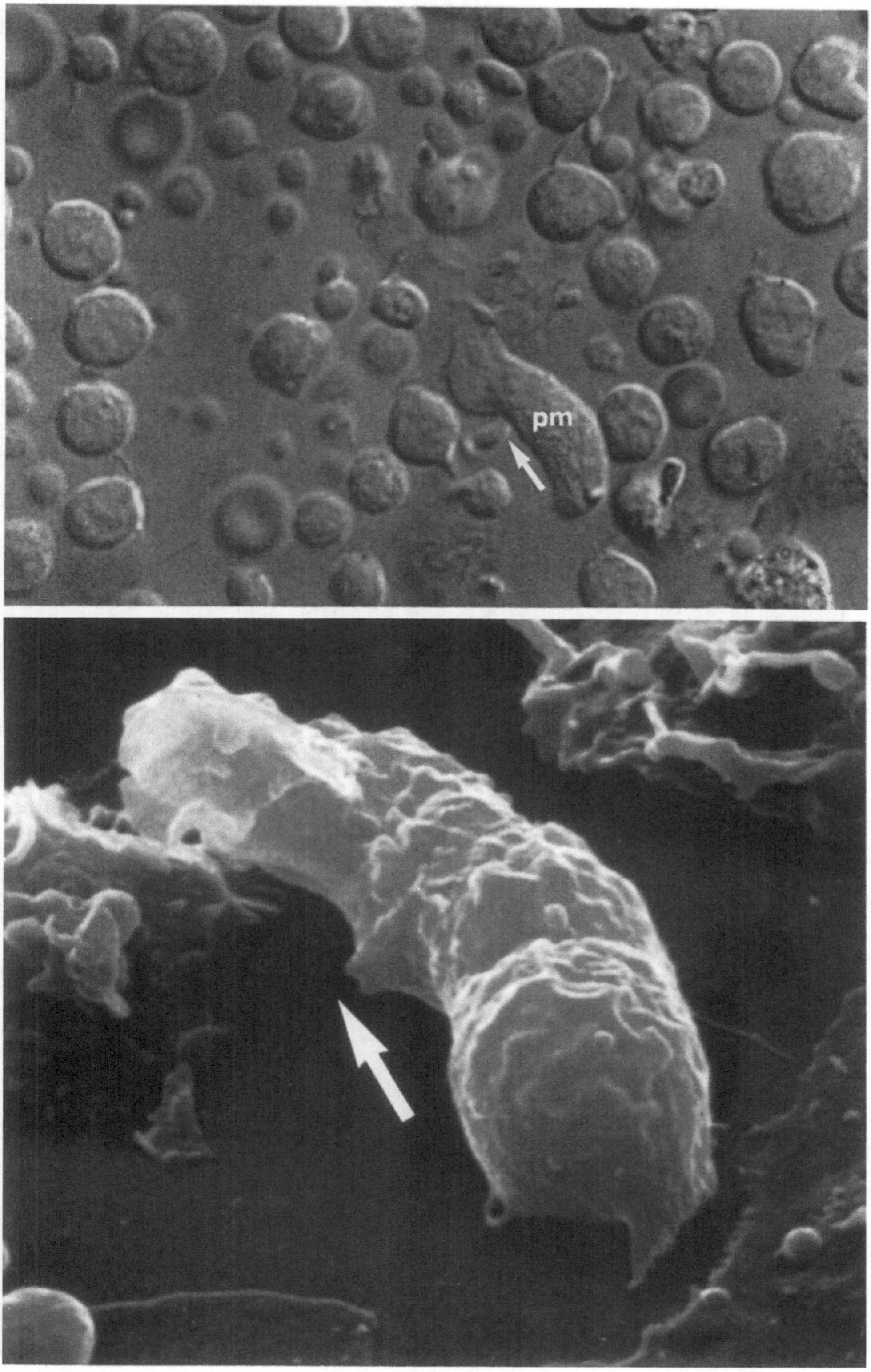

a

b

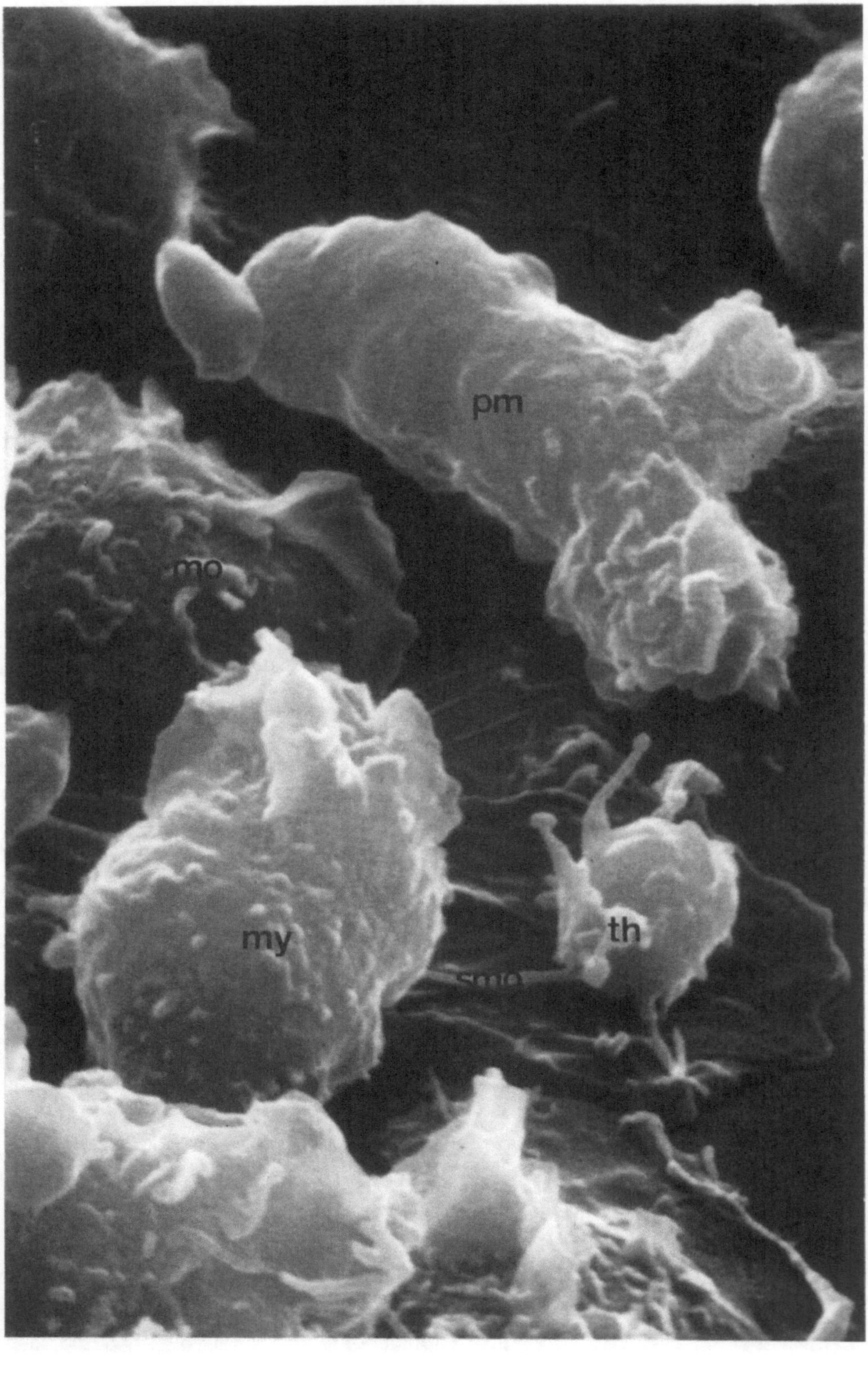
pm
mo
my
th
smo

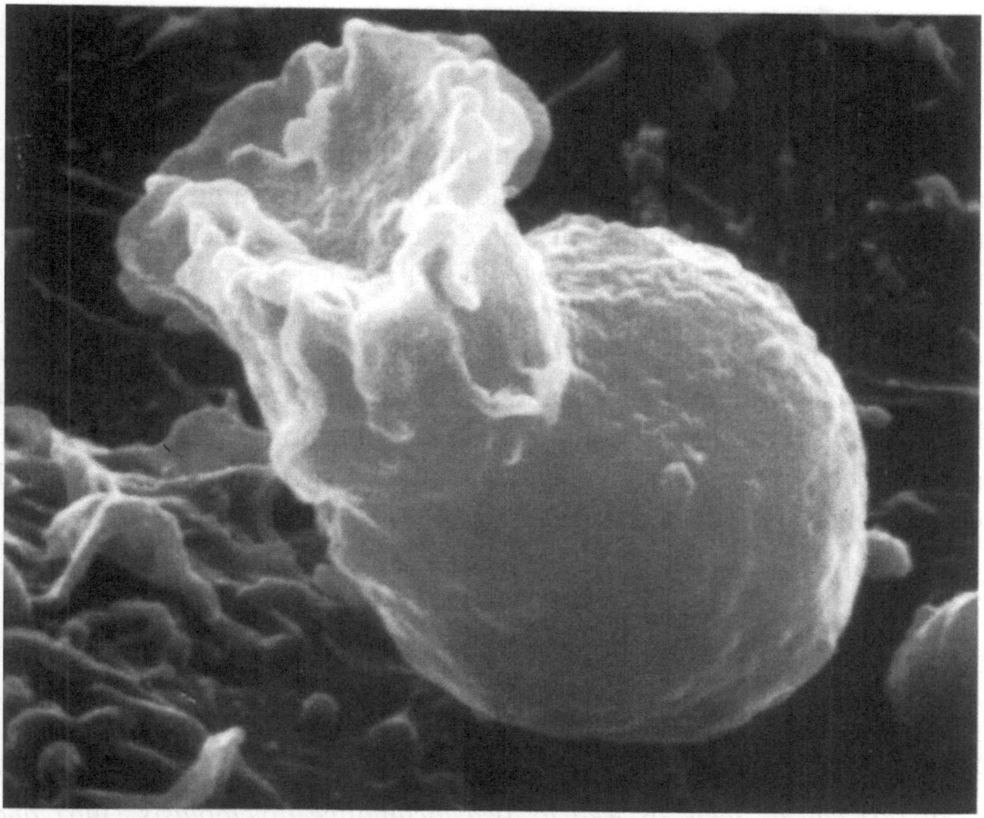
a

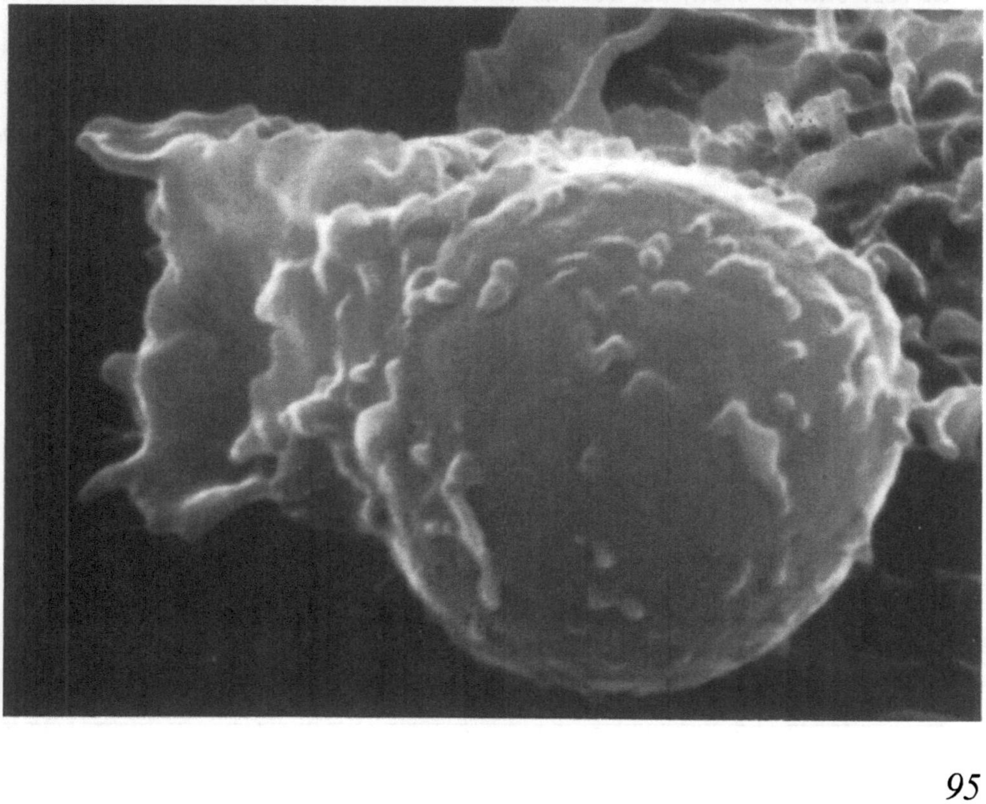
b

This again is a heterogenous population with myeloblasts and monocytes as the main components, while the more differentiated types of the myeloid series are less numerous. Before onset of treatment the locomotive activity is restricted to myeloid cells, whereas monocytes show surface motility only. After the patient has been treated for 5 months, cellular composition and behavior are changed. The population now mainly consists of monocytes and the more mature forms of myeloid cells. Analysis of time-lapse films shows that the monocytes have become slowly locomotive. These changes are documented by both SEM and cinematography.

Figure 56. Leukemic population (before treatment)

a In the phase-contrast photograph from a time-lapse film various cells at rest can be seen, as well as two locomotive myeloblasts (mb) and a locomotive myelocyte (my)

b SEM shows a spread monocyte (mo), a polarized myeloblast (mb), a myelocyte (my) that could be locomotive, and a spheric cell of unknown origin

a Phase contrast: × 2100
b SEM: × 7400

a

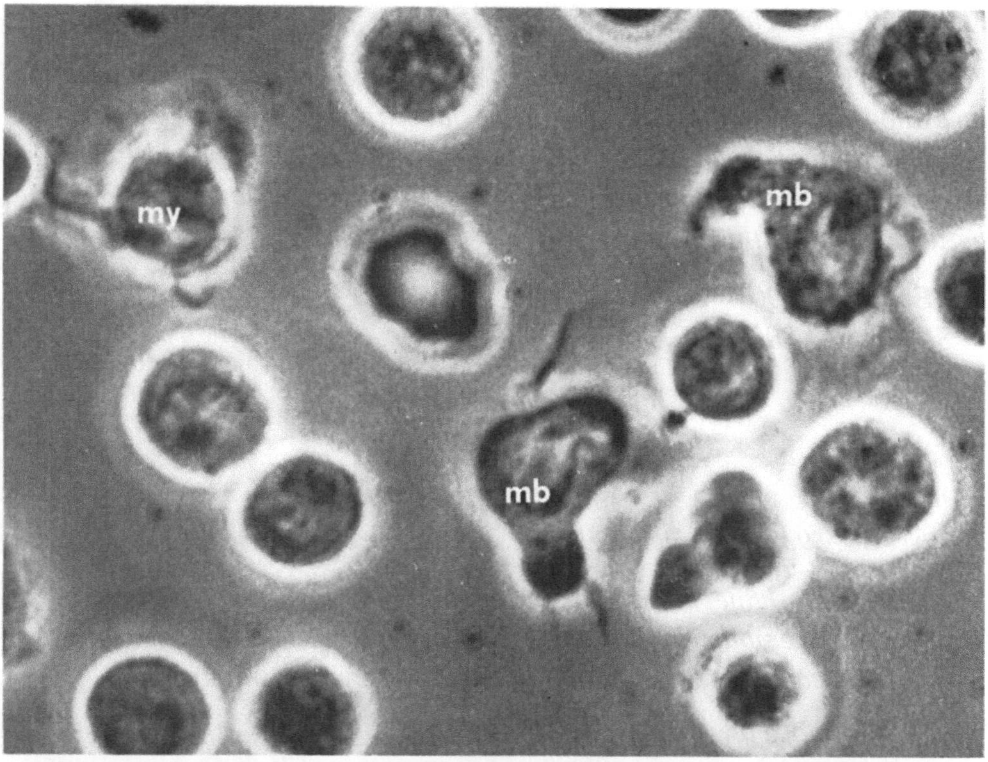

b

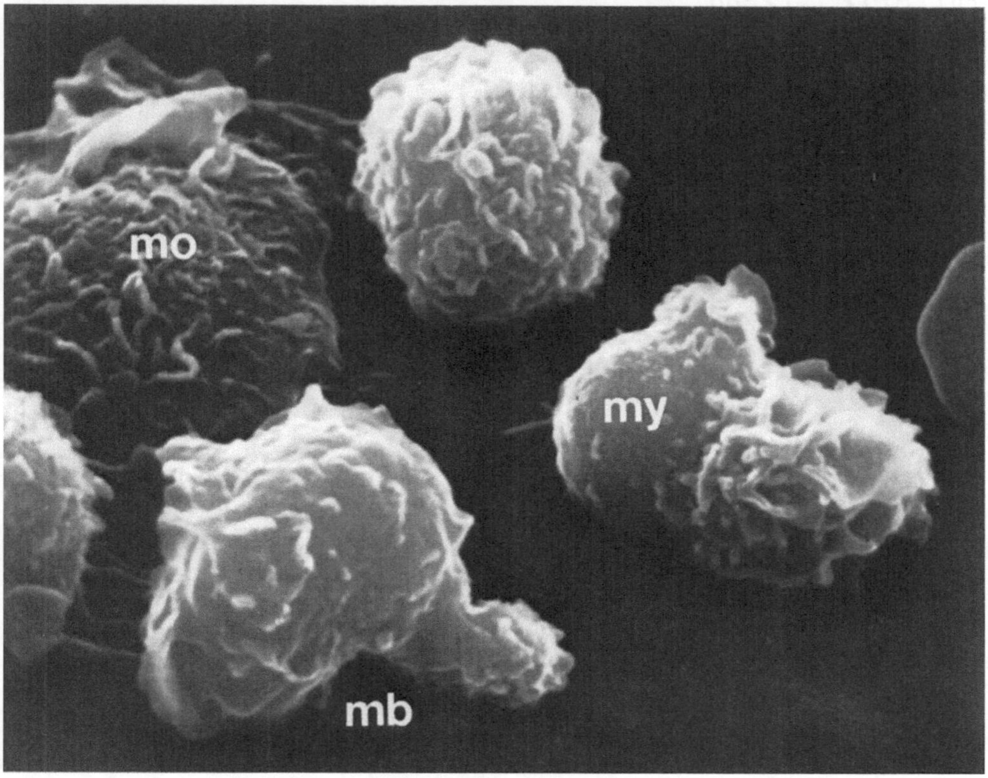

Figure 57. Leukemic cells (before treatment)

Two spheric cells on a spread monocyte and one on glass. Surface architecture alone does not permit definitive identification of these cells. The configuration of the polarized cell, on the other hand, is typical for blasts. Whether this cell is performing on-spot motility or locomotion cannot be decided from this picture

a SEM: × 7500
b SEM: × 14,800

Figures 58, 59. Leukemic myelocytes (before treatment) *(see pages 100, 101)*

The range of shape variations within the configurational pattern of the myelocyte is demonstrated in these SEM pictures. The direction of movement is indicated by the position of the large cytoplasmic veils

58a SEM: × 13,900
58b SEM: × 14,200
59a SEM: × 15,300
59b SEM: × 18,500

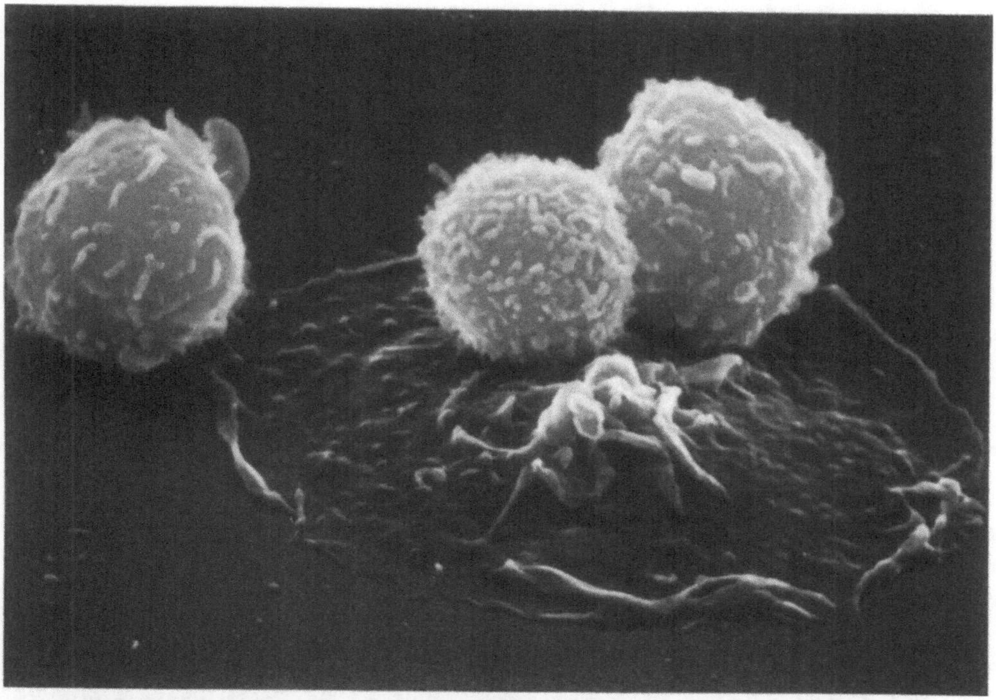

a

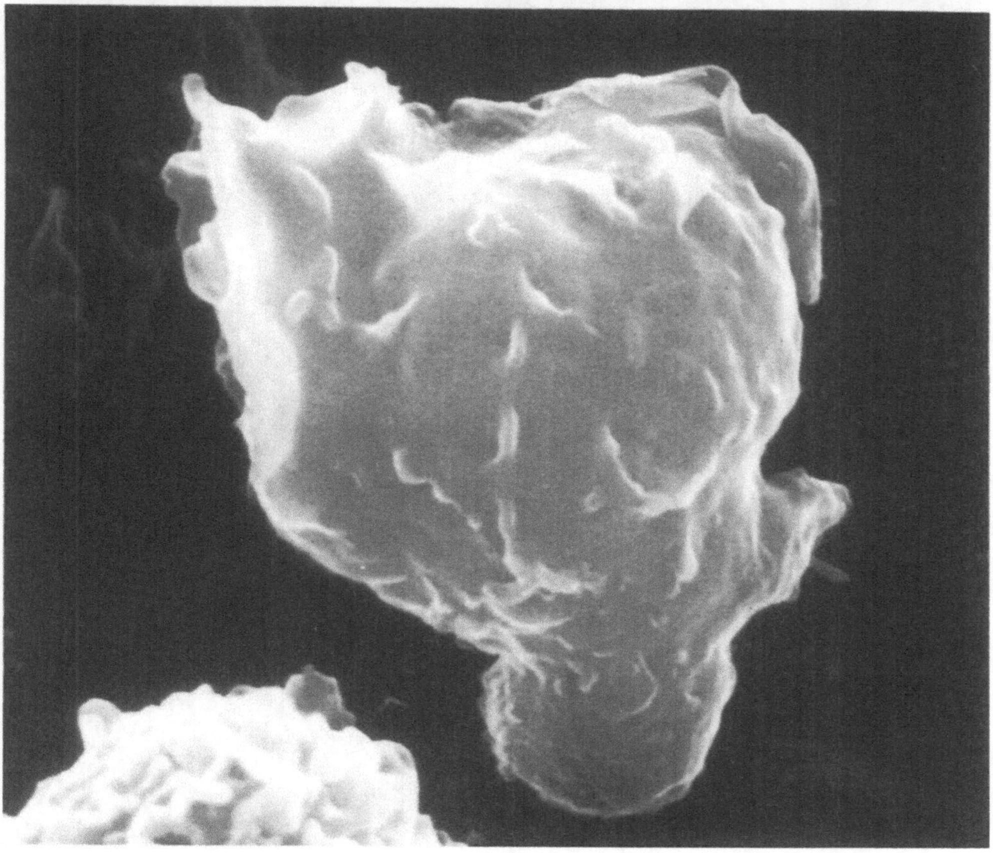

b

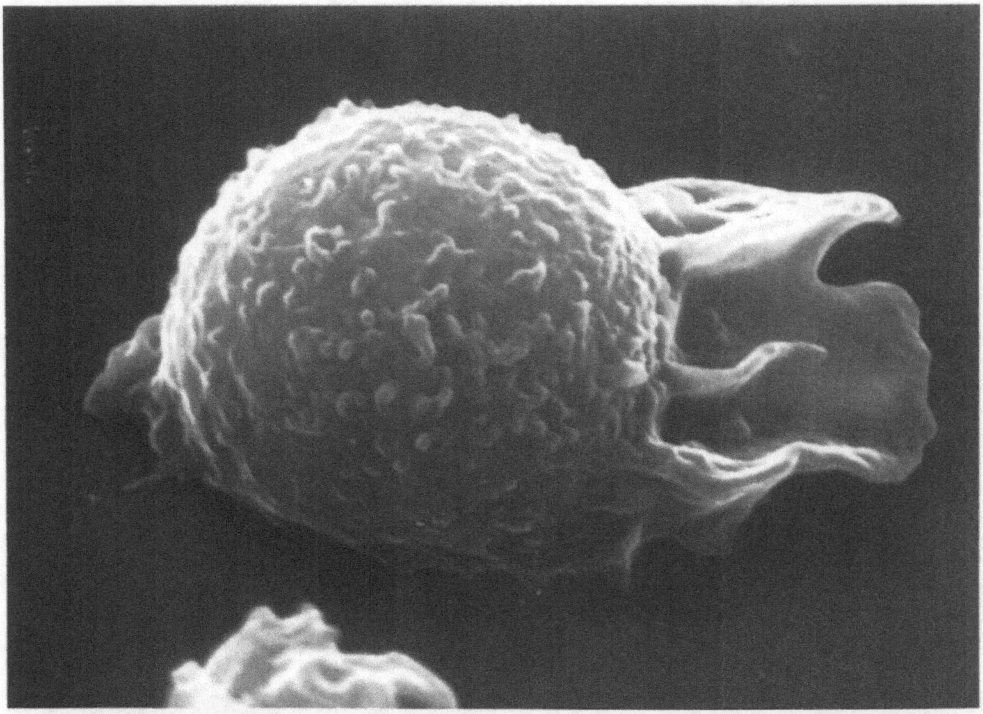

a

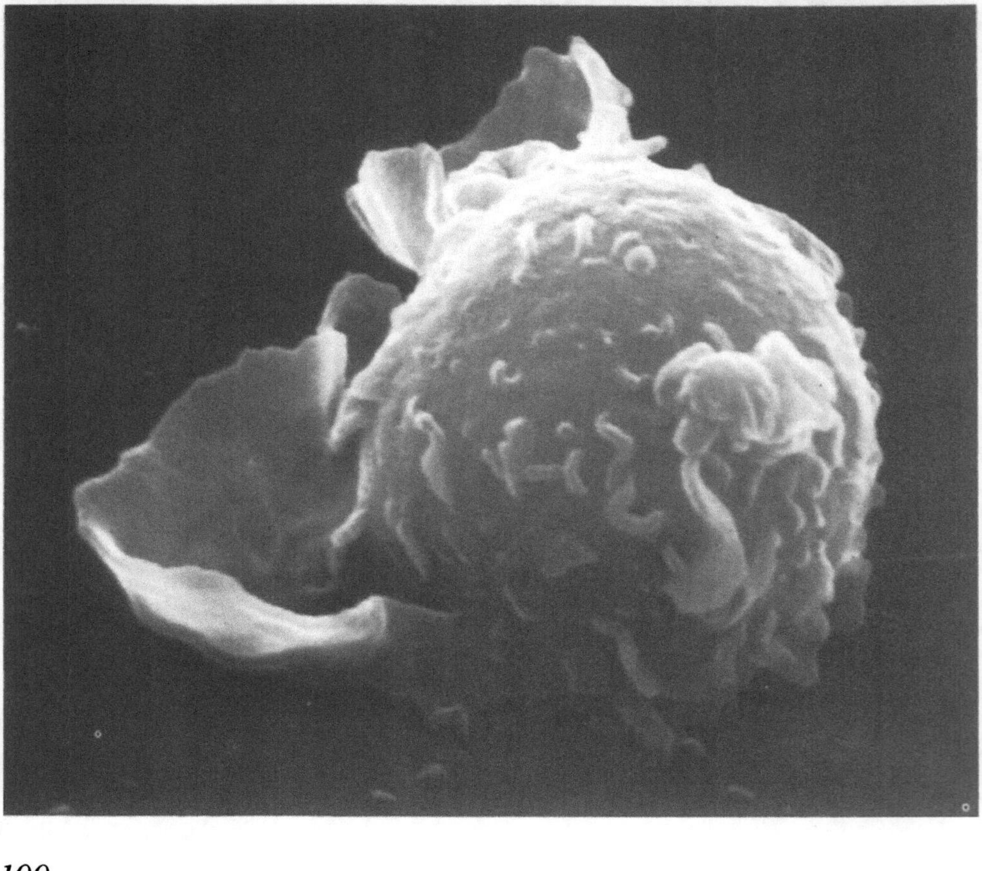

b

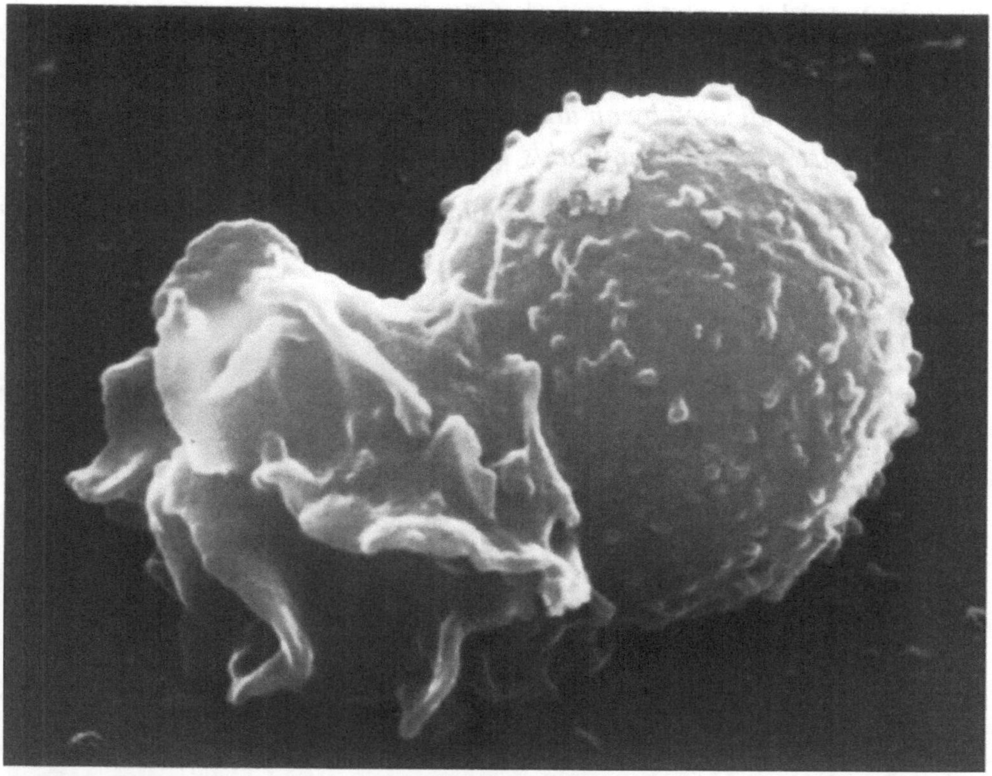

a

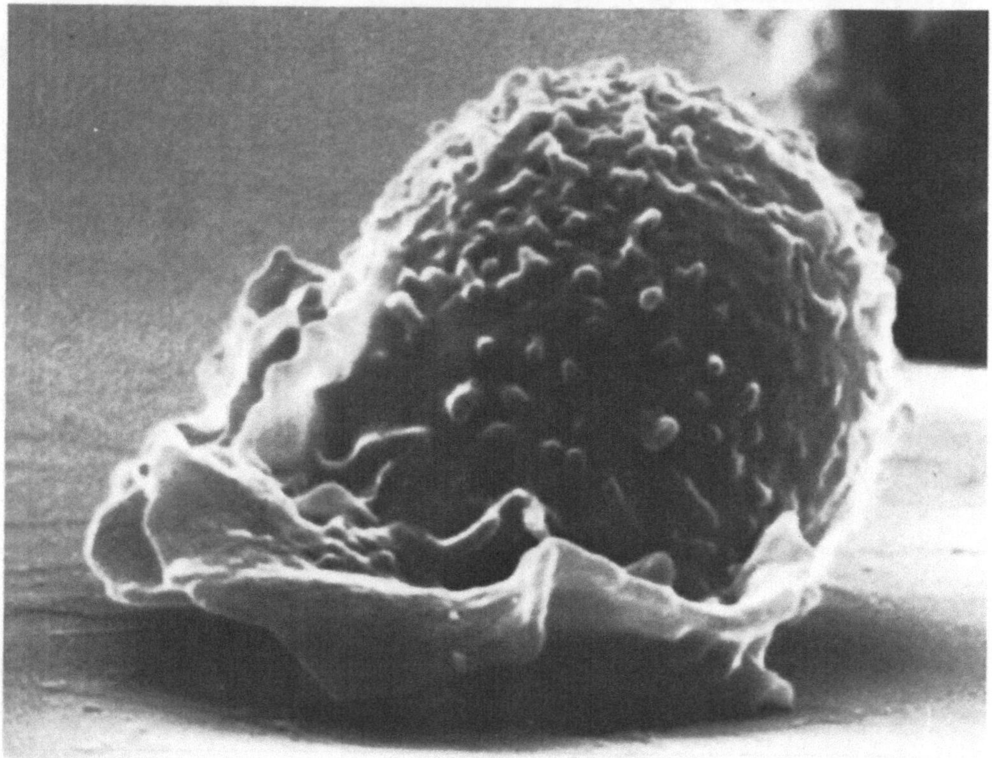

b

Figure 60. Leukemic monocytes (before treatment)

a The characteristic aspect of a monocyte fixed during the process of spreading is indicated by an already flattened area around a still spheric central part with the many folds typical for monocytes

b Well-spread monocytes without any suggestion of locomotive activity

a SEM: × 19,000
b SEM: × 3400

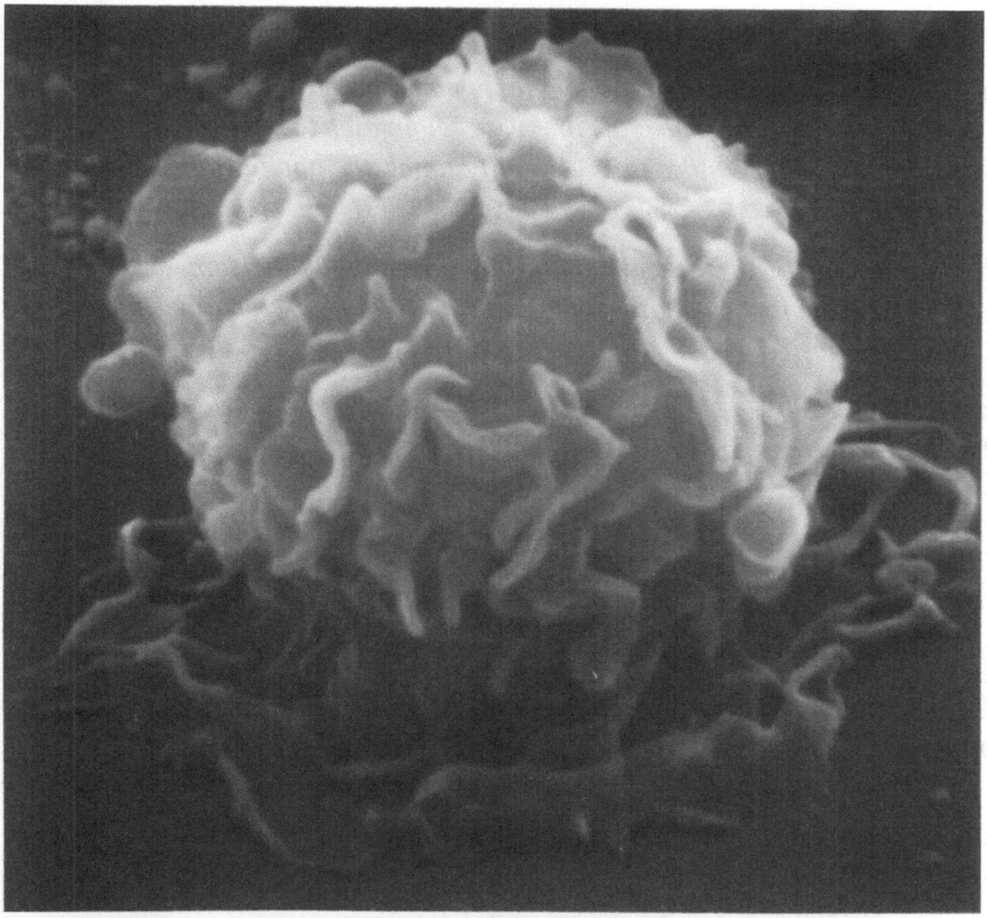

a

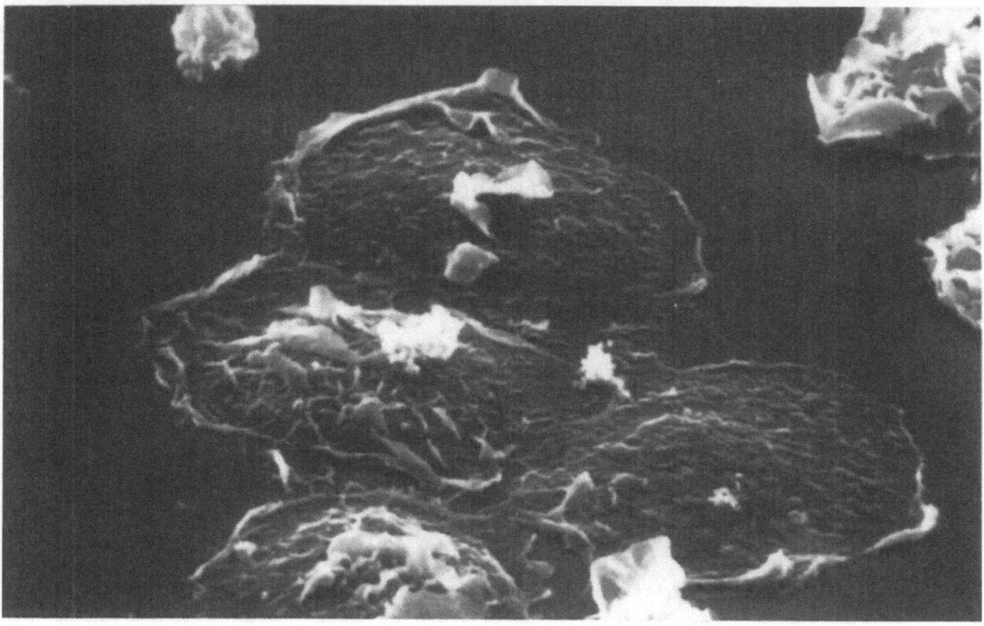

b

Figure 61. Leukemic population (during treatment)

a Conspicuous among the monocytes (mo), spread to a lesser or greater extent, is the presence of the very elongated cell with a long tail. Monocytes shaped like this were not among the population before treatment

b The similarity to the cell in the photograph taken from a time-lapse film is evident. Cinematography reveals that such elongated monocytes are locomotive

a SEM: × 3500
b Phase contrast: × 1200

Figure 62. Leukemic population (during treatment) *(see page 106)*

Monocytes in various shapes are conspicuous during the period of treatment

SEM: × 3800

Figure 63. Leukemic monocytes (during treatment) *(see page 107)*

The SEM picture shows a spheric and a polarized monocyte with extensive veils at the anterior part (a), and another polarized locomotive monocyte together with thrombocytes, spread and nonspread

a SEM: × 7300
b SEM: × 6900

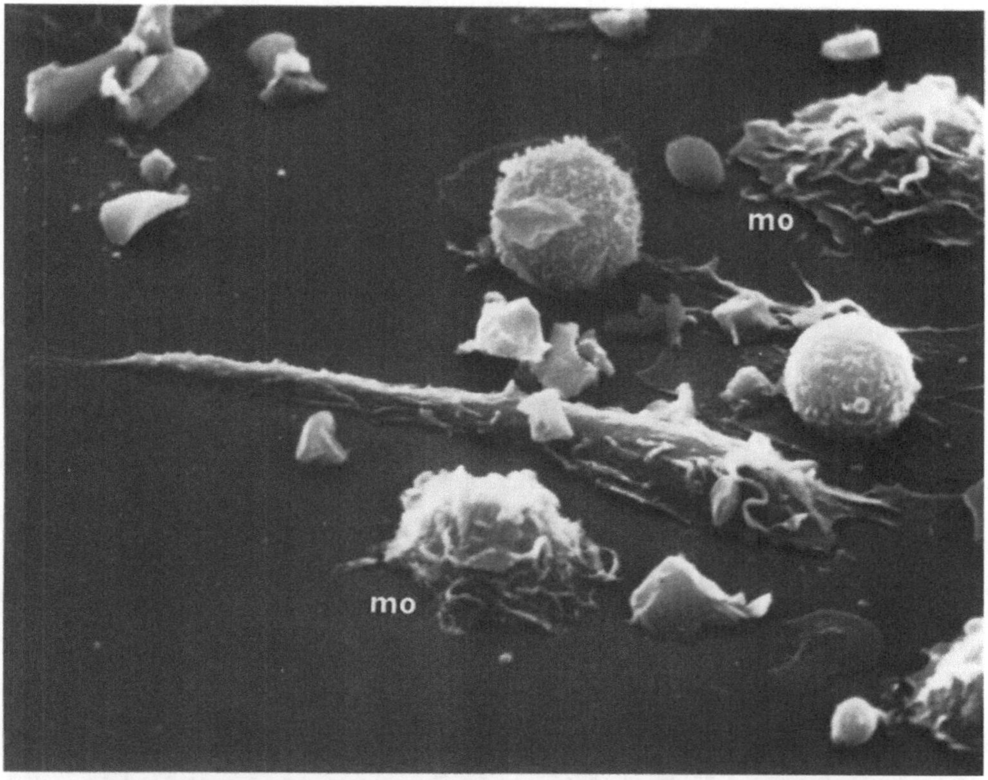

a

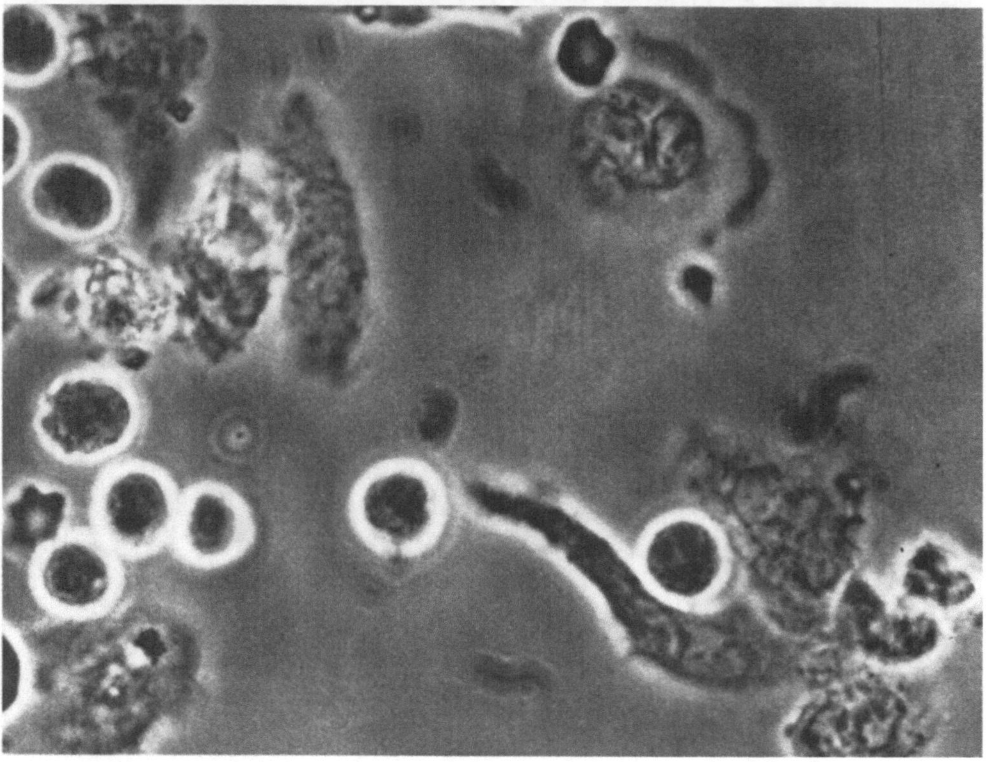

b

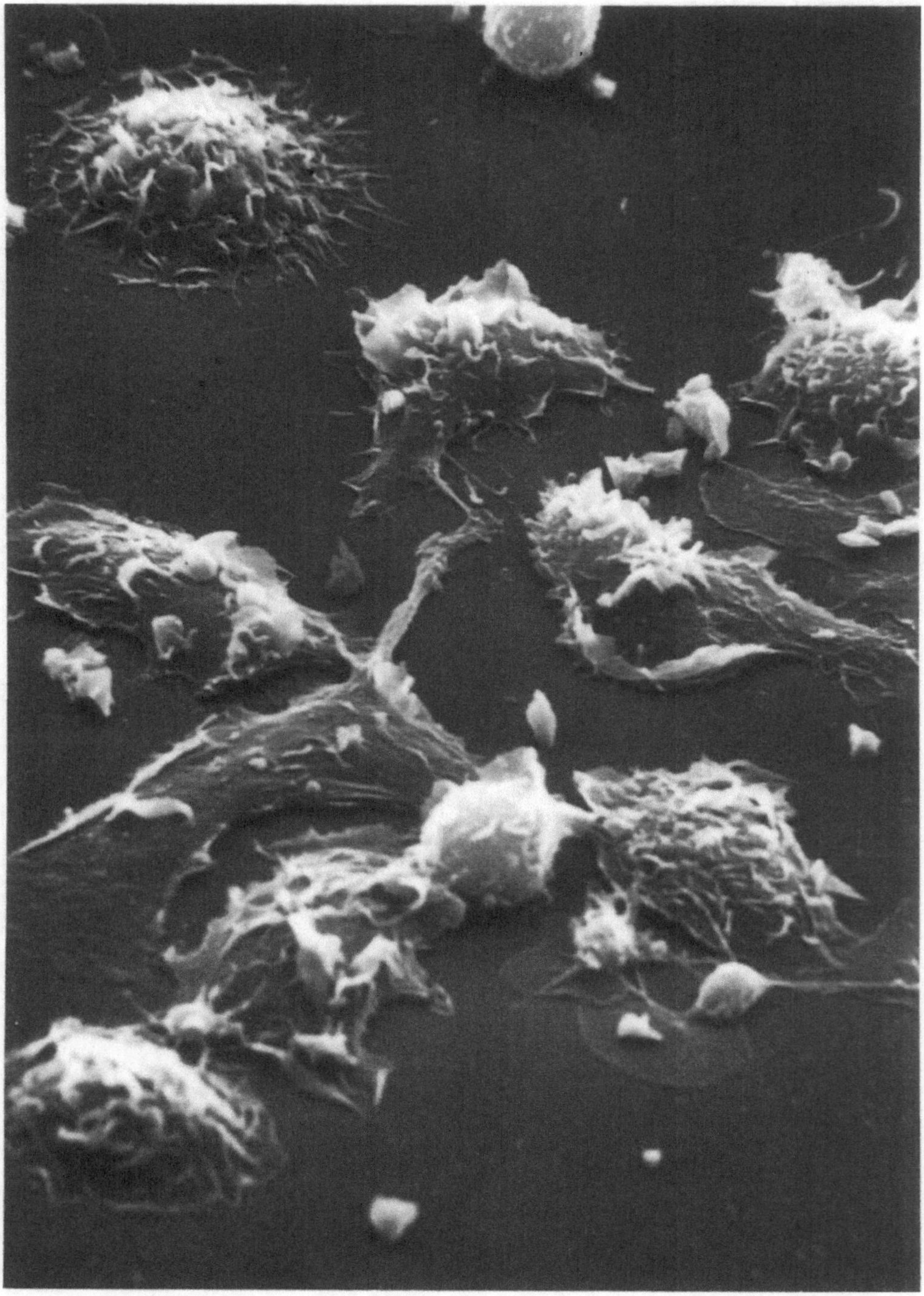

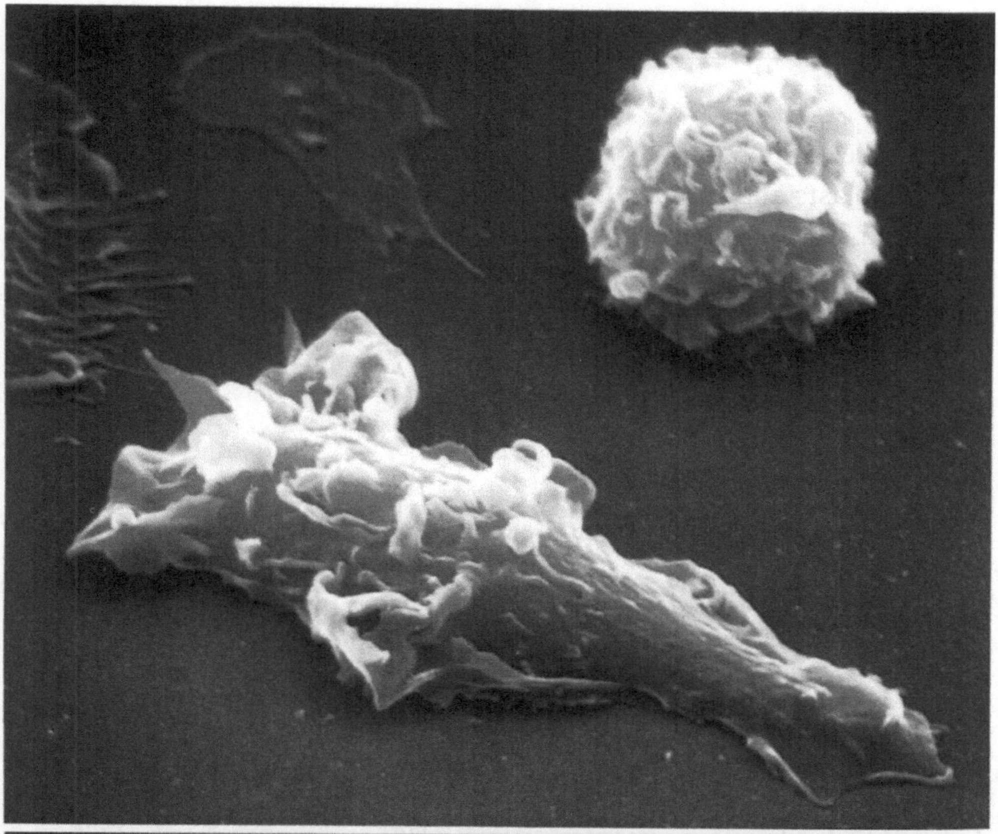

a

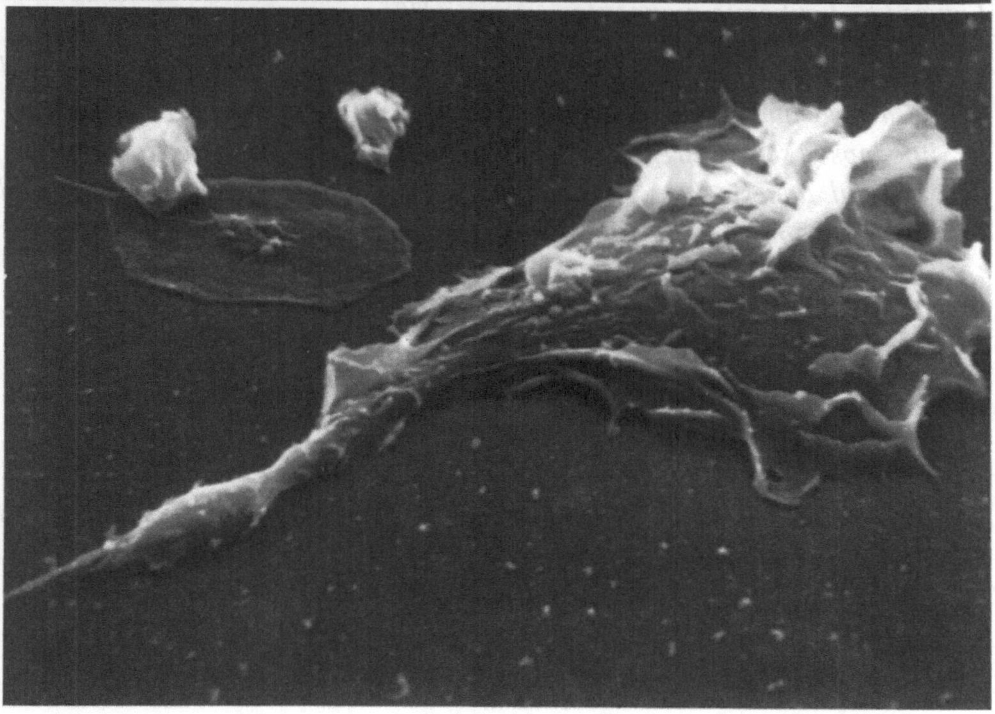

b

Again a leukemia in which the majority of the population consists of myeloblasts and monocytes. The monocytes in our preparation have elongated bodies and show the tendency of forming clumps. They are slowly locomotive. On the other hand, the activity of most of the myeloblasts is restricted to on-spot motility; slow locomotion of short duration is infrequent.

Figure 64. Leukemic population

The long elongated shape of the monocytes is complementarily documented by SEM (a) and phase contrast (b). The two photographs also show a cell clump which is blurred in phase contrast

a SEM: × 3000
b Phase contrast: × 1200

Figure 65. Leukemic monocytes *(see page 110)*

These are two monocytes, one partially spread (a), the other elongated (b) with large and flat cytoplasmic extensions at the front part. As verified by time-lapse (see Figure 64b) the elongated position indicates locomotion. The special adhesiveness of the long cell to the substrate is suggested by the appearance of the flat, irregularly shaped posterior extension (b)

a SEM: × 8600
b SEM: × 7000

Figure 66. Leukemic monocytes *(see page 111)*

The special configuration of the monocytes of this leukemia is emphasized by these SEM photographs

a SEM: × 4600
b SEM: × 7300

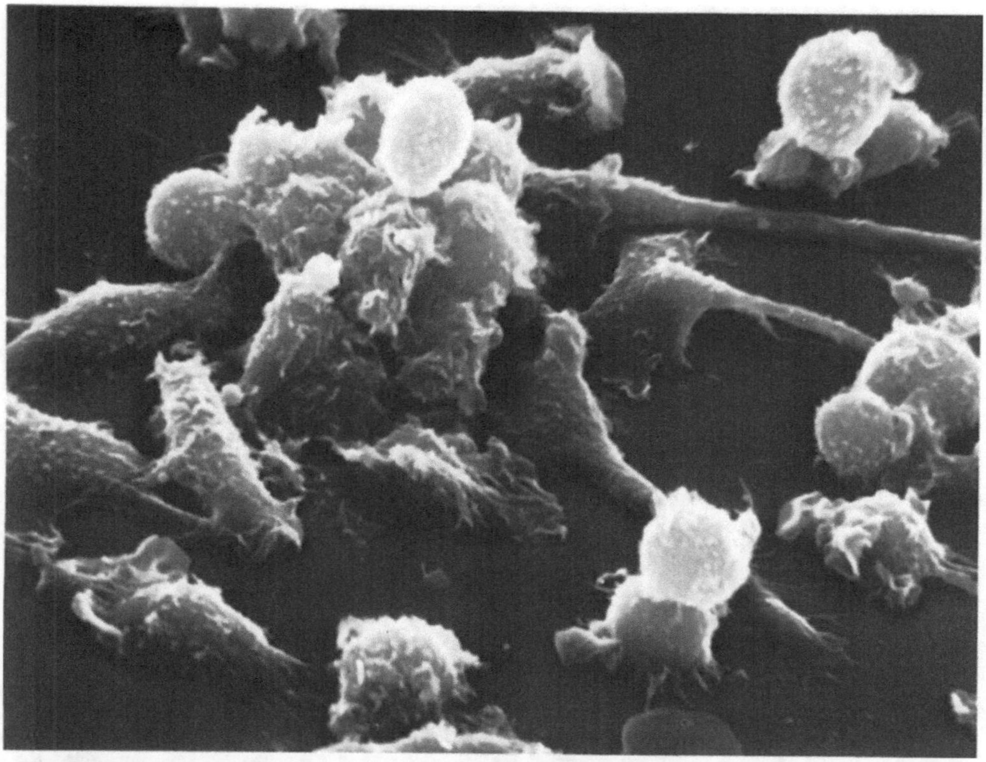
a
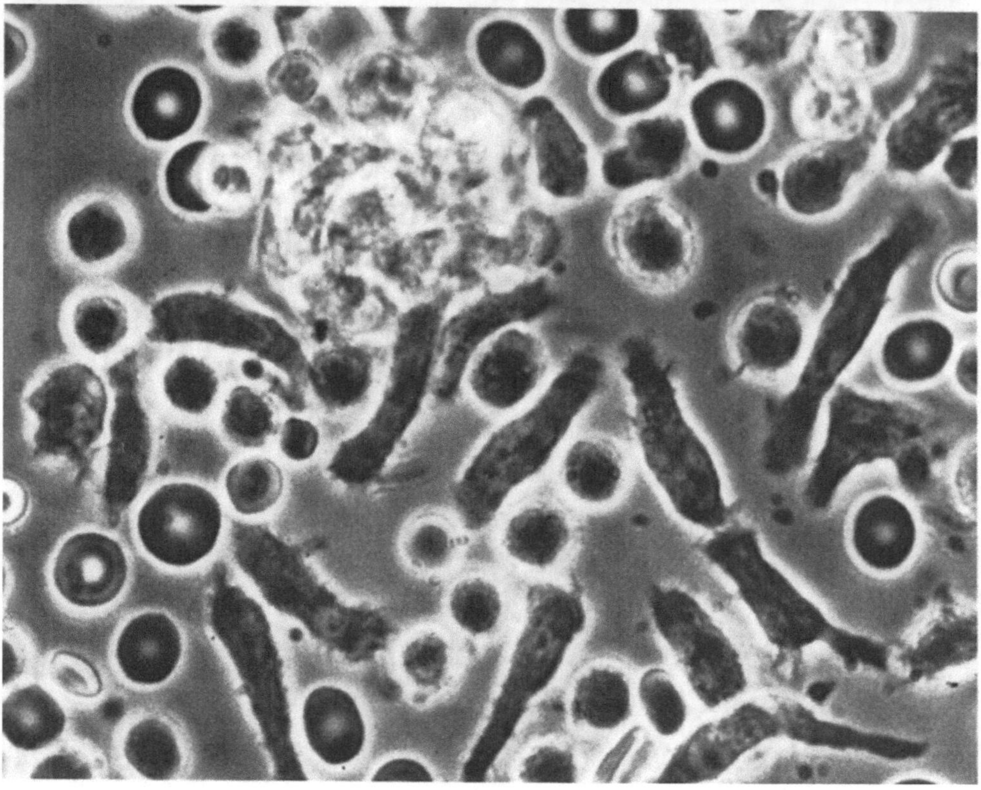
b

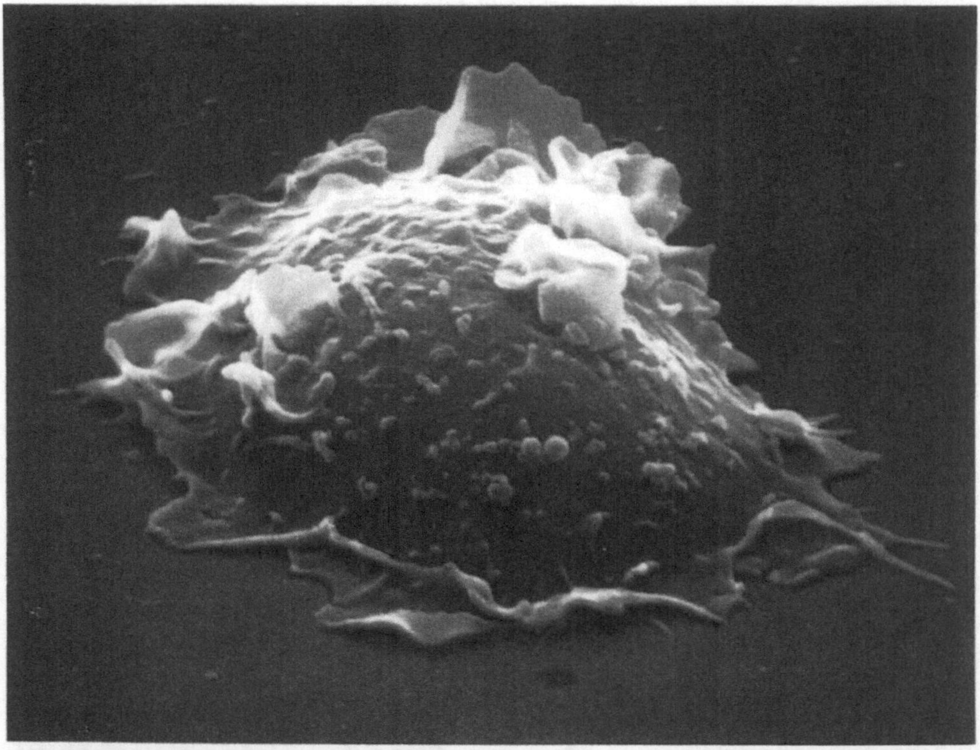
a

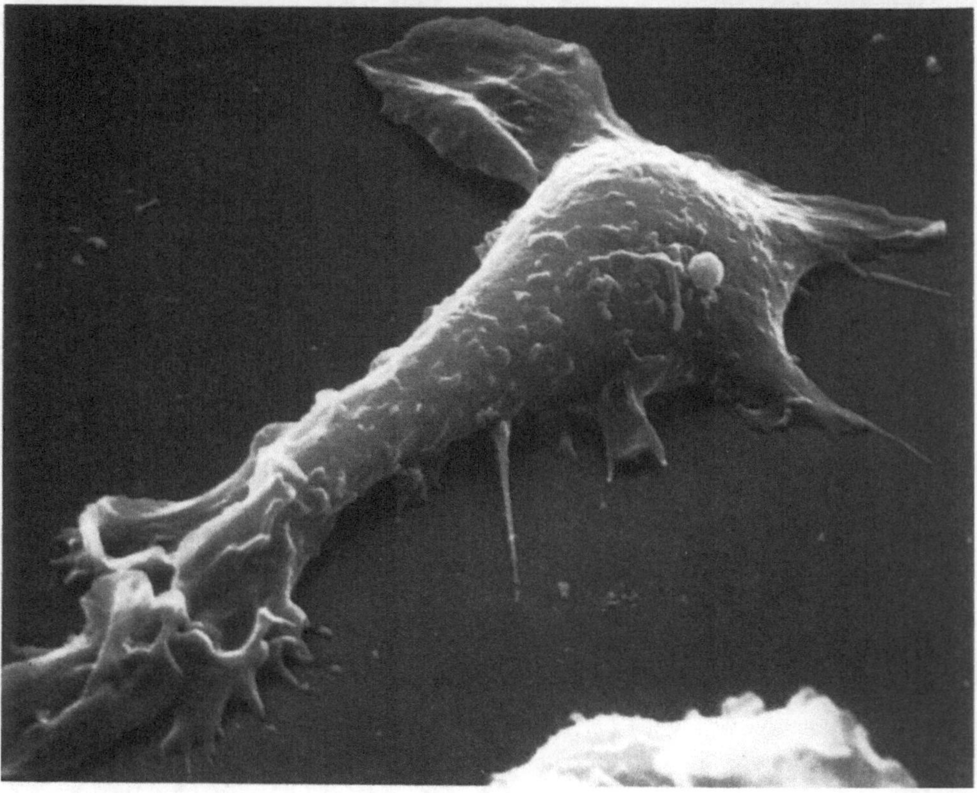
b

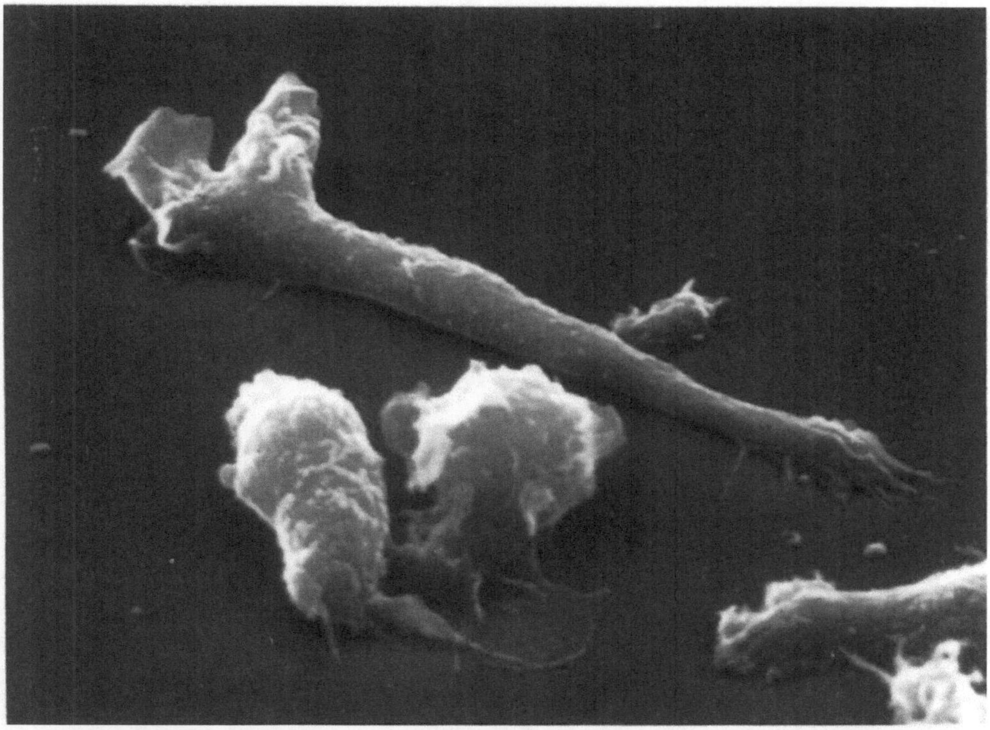

a

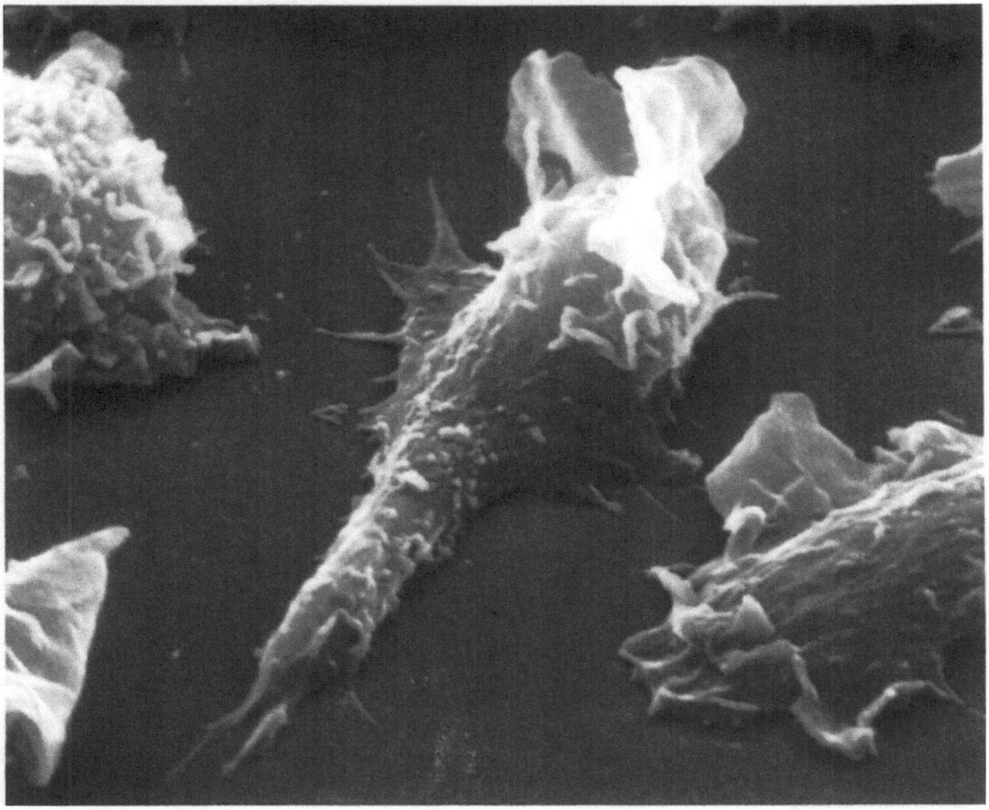

b

Figure 67. Leukemic monocytes

Underlapping? Overcrossing? This cannot be decided on fixed preparations. For some of the monocytes, both possibilities seem to occur (b)

a SEM: × 7200
b SEM: × 7000

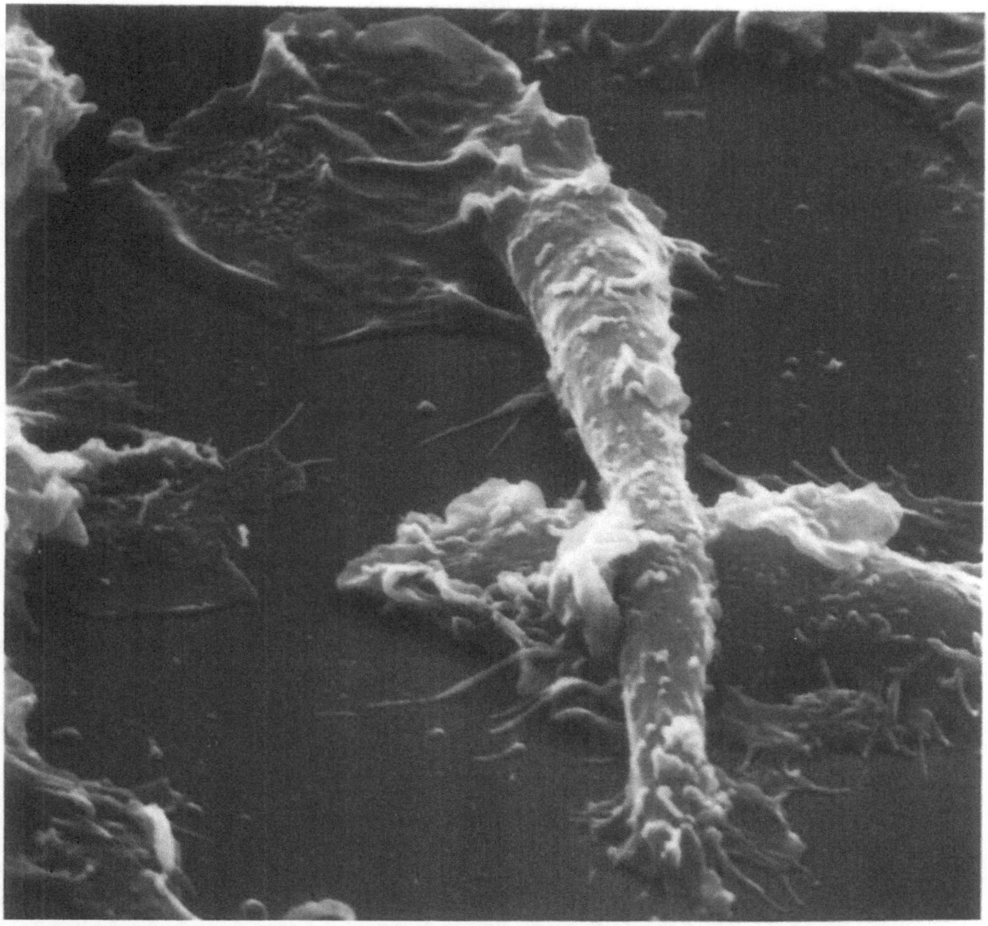

a

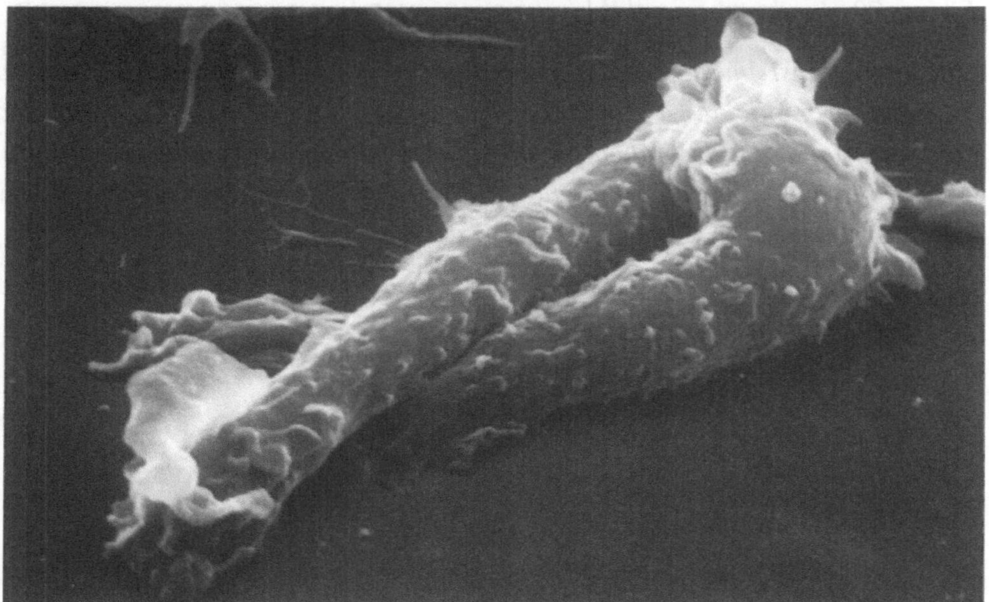

b

Figure 68. Leukemic population

This cell clump (a) contains two types of spheric cells with different surface structures. Those with the large, irregular folds are monocytes (mo). The cells with the regular pattern of short microvilli (a, b) cannot be classified

a SEM: × 3400
b SEM: × 14,600

Figure 69. Leukemic myeloblast *(see page 116)*

The polarized configuration suggests the blast nature of this cell that is partly adherent to the glass surface, partly to the edge of a spread monocyte. Shape and surface extensions at the top of the cell are indicative of on-spot motility

SEM: × 21,000

Figure 70. Leukemic myeloblasts *(see page 117)*

One myeloblast in the polarized configuration is fastened by its foot to a spread thrombocyte (a), the other (b) to the glass surface. The polarized configuration could be indicative of both on-spot motility and locomotion. The position of the cells does not allow to decide during which activity they were fixed

a SEM: × 14,600
b SEM: × 15,500

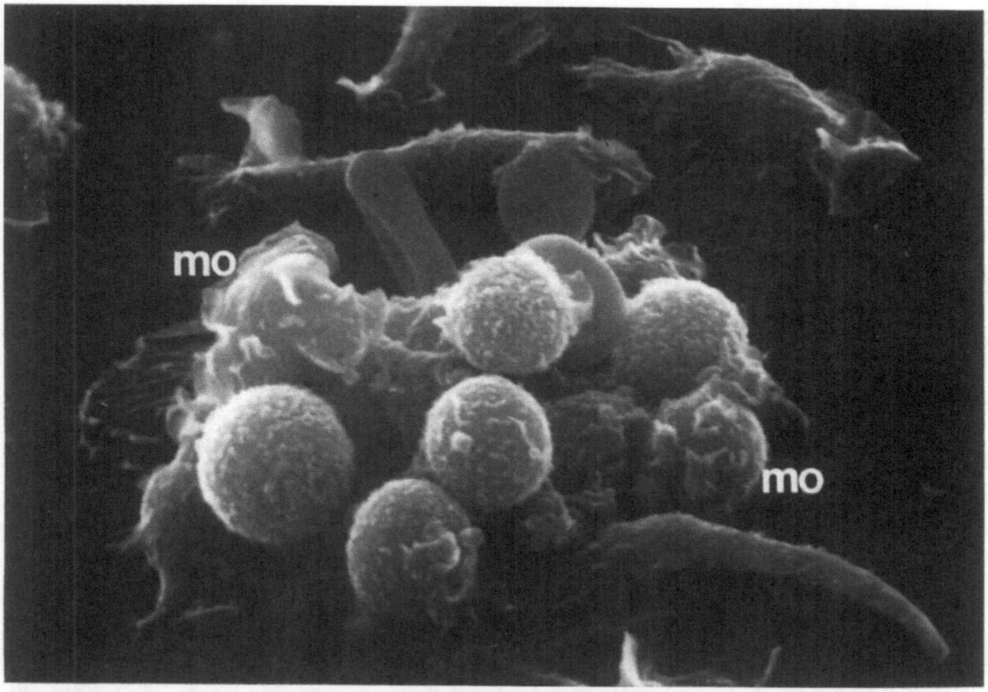

a

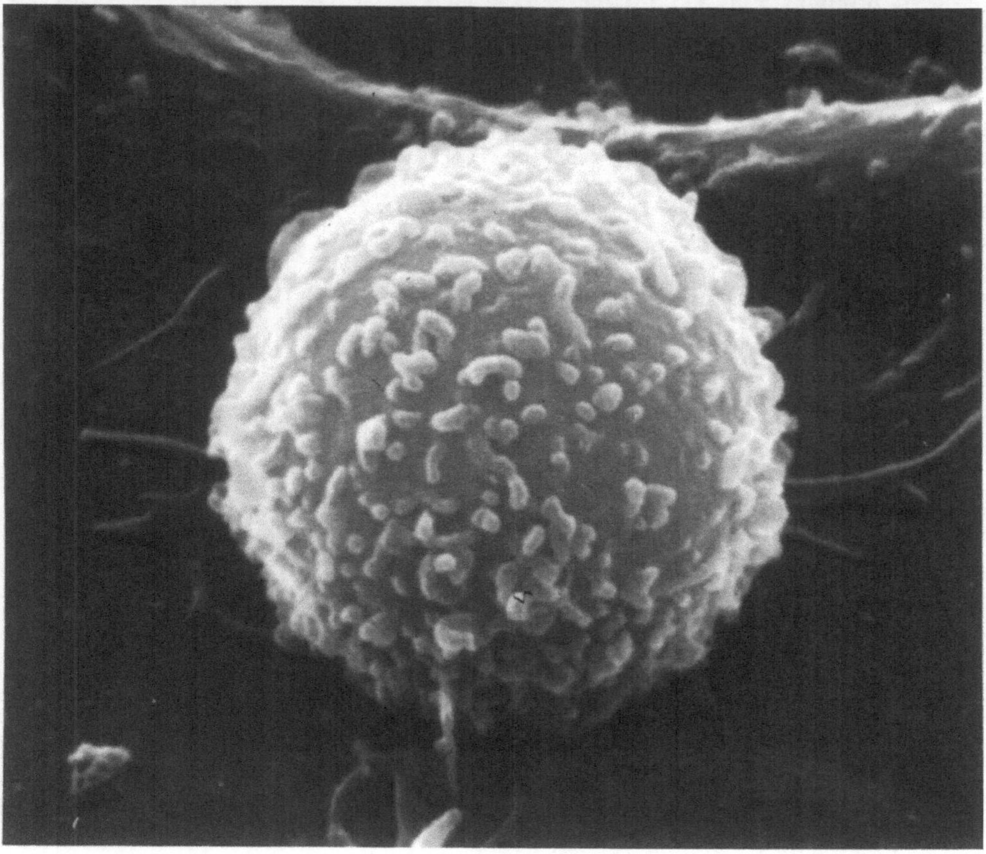

b

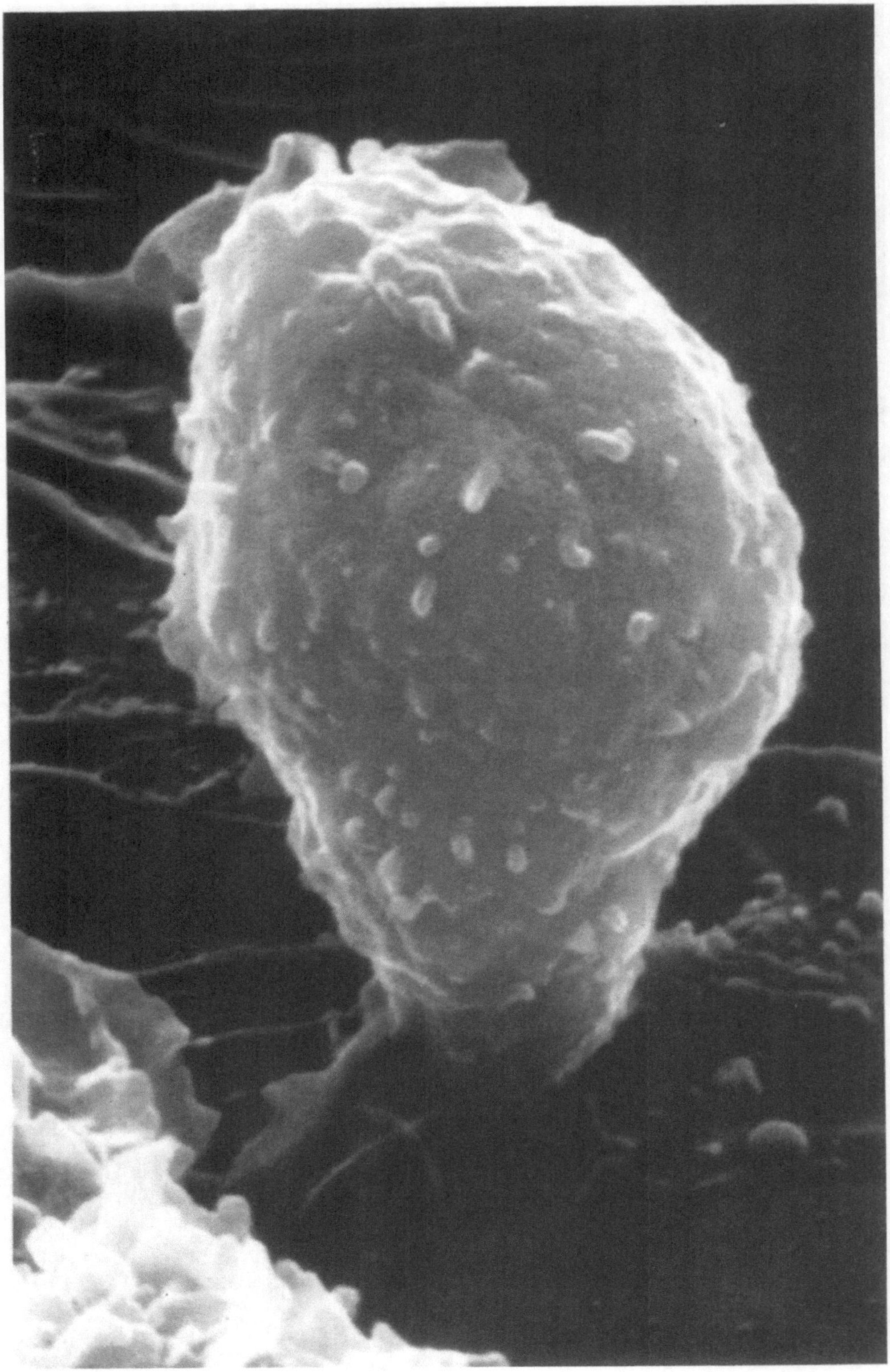

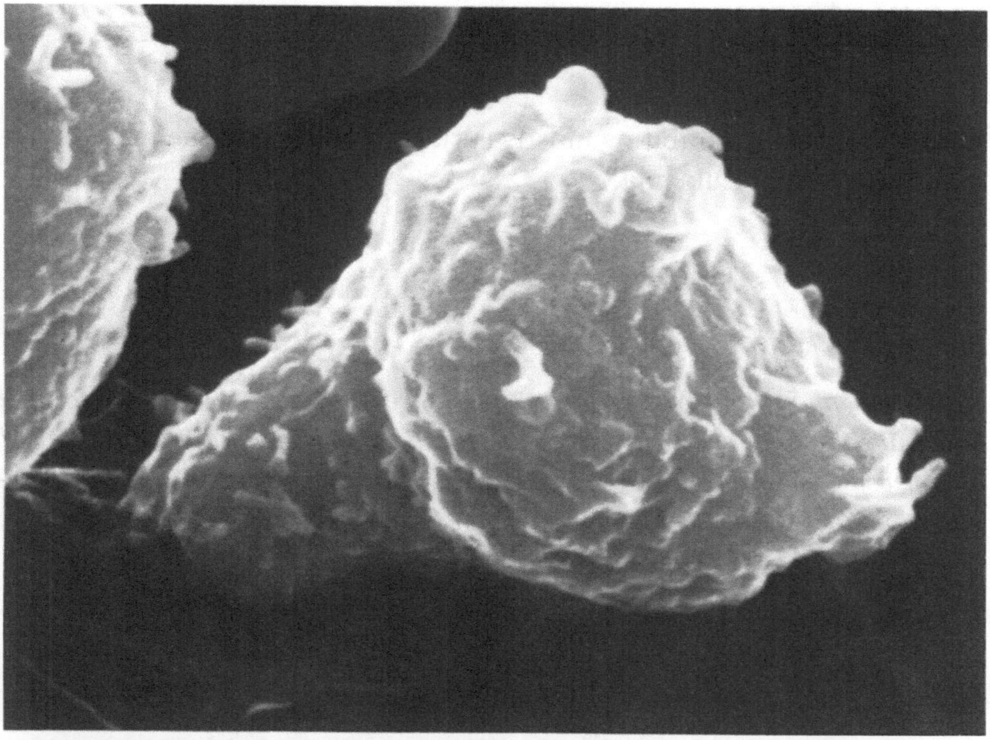
a

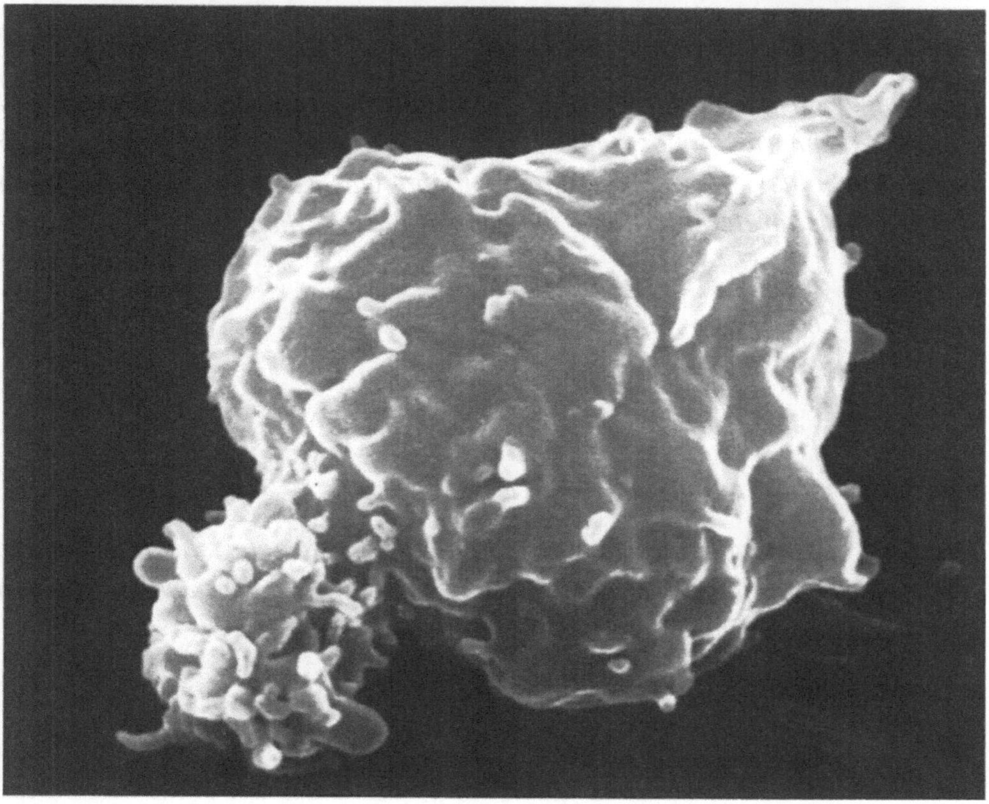
b

Monocytoid cells, exhibiting mostly surface motility, are the main representatives of this leukemia.

Figure 71. Leukemic population

SEM and phase contrast from a time-lapse film show that most of the leukemic monocytoid cells are spread in various configurations. SEM also reveals the presence of spheric cells. Note the similarity of the elongated monocytes in the two photographs (arrows)

a SEM: × 3200
b Phase contrast: × 2200

Figure 72. Leukemic monocytoid cells *(see page 120)*

The typical surface architecture of monocytes is presented in these photographs. A spheric monocyte (a) is covered in a regular manner with large, thin folds, whereas the spread monocytes (a, b) can retain some of these folds, preferentially in the central part of their body

a SEM: × 7800
b SEM: × 7600

Figure 73. Leukemic monocytoid cells *(see page 121)*

These spread monocytes have also kept their characteristic folds in the center (a). The flatness of the cells and the bulge of the nucleus, covered with folds, is especially well seen at a tilt angle of 85° (b)

a SEM: × 8100
b SEM: × 9900

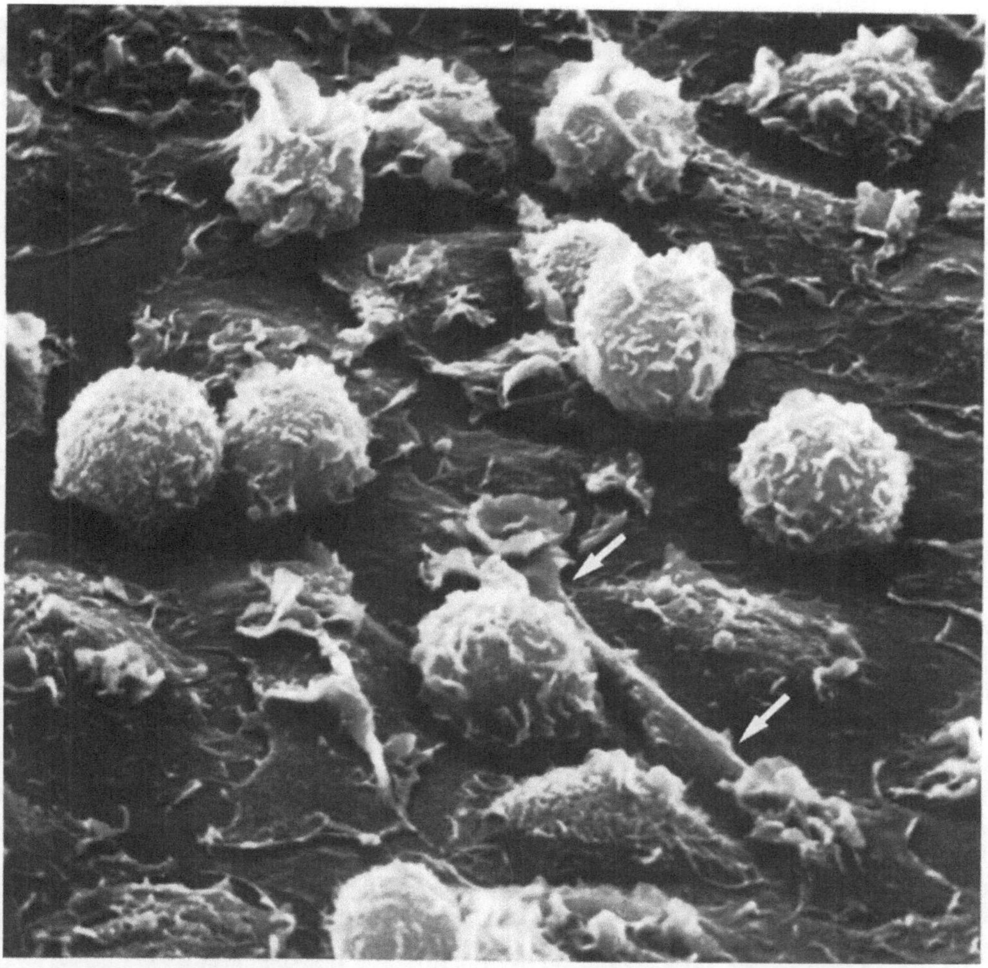
a

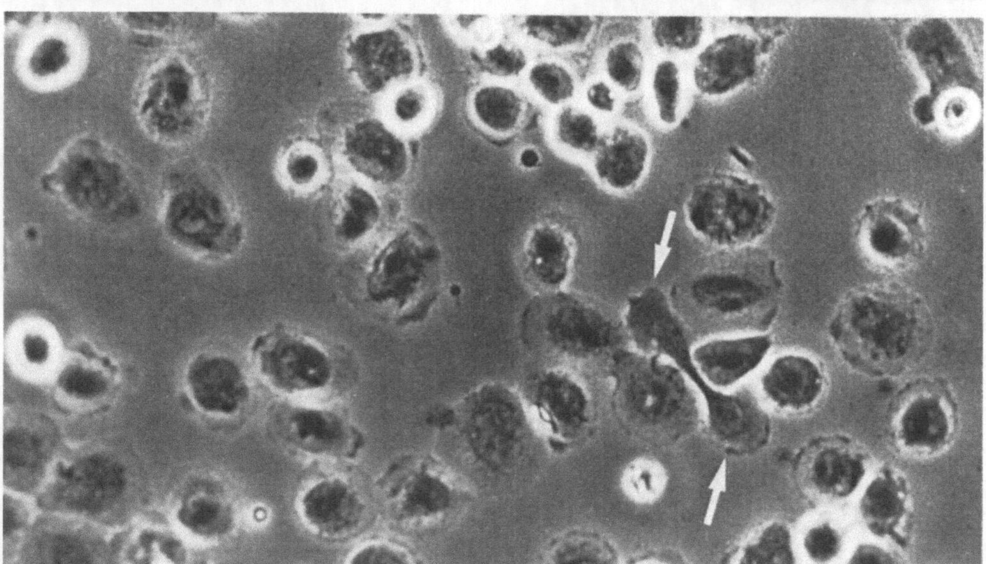
b

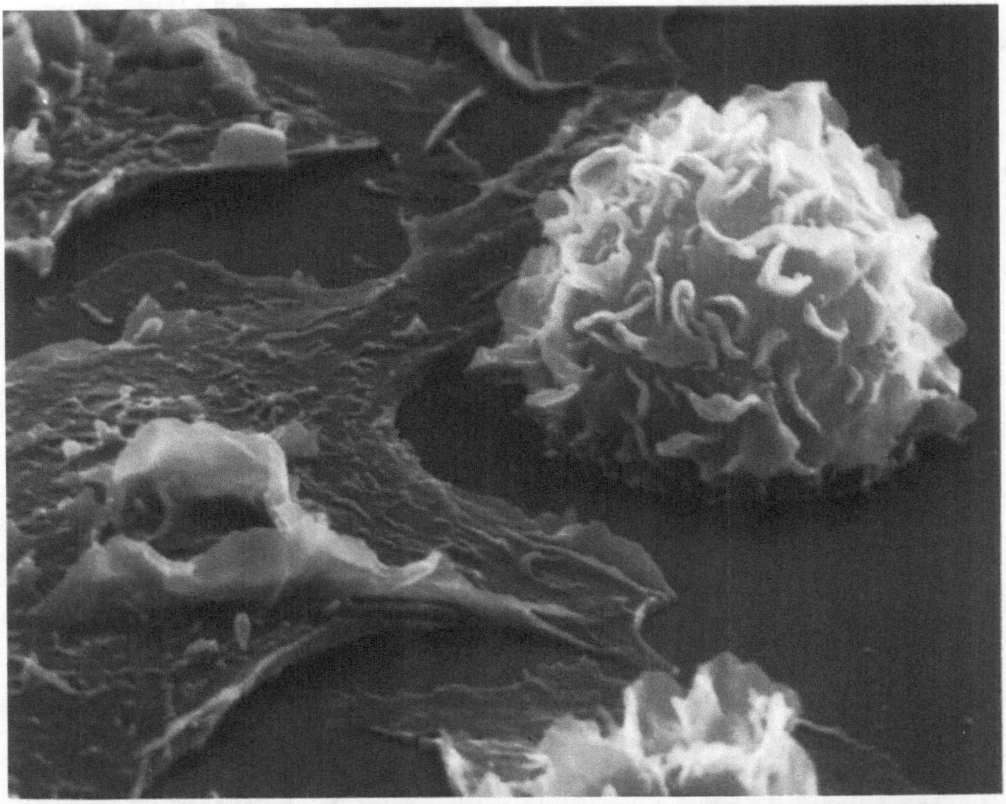

a

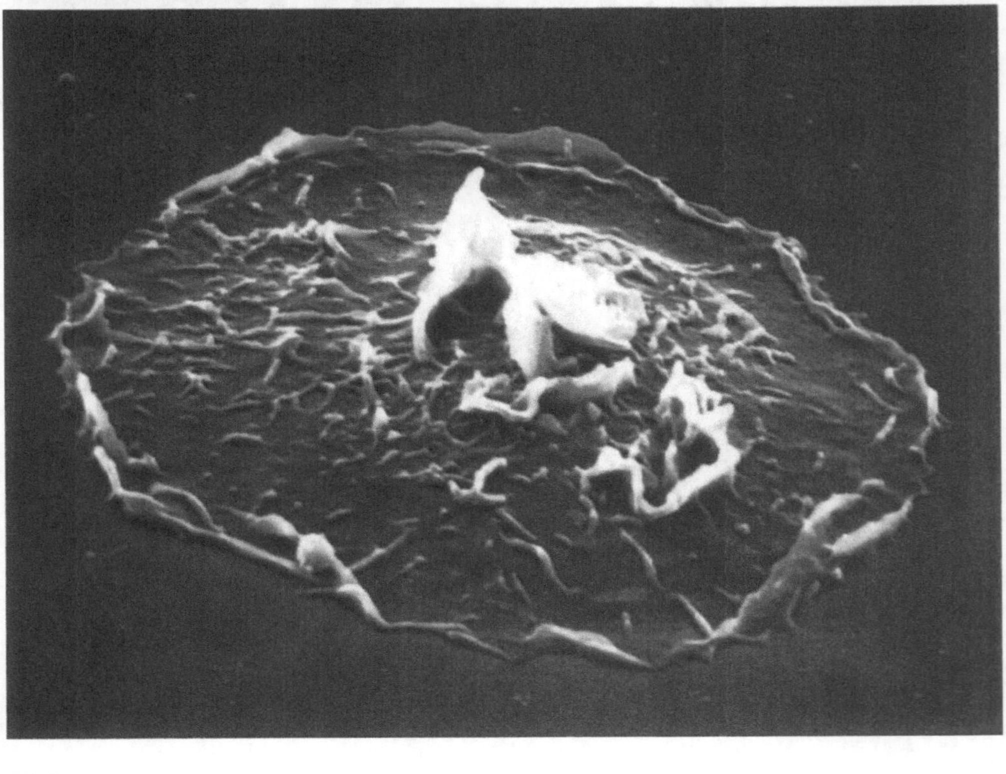

b

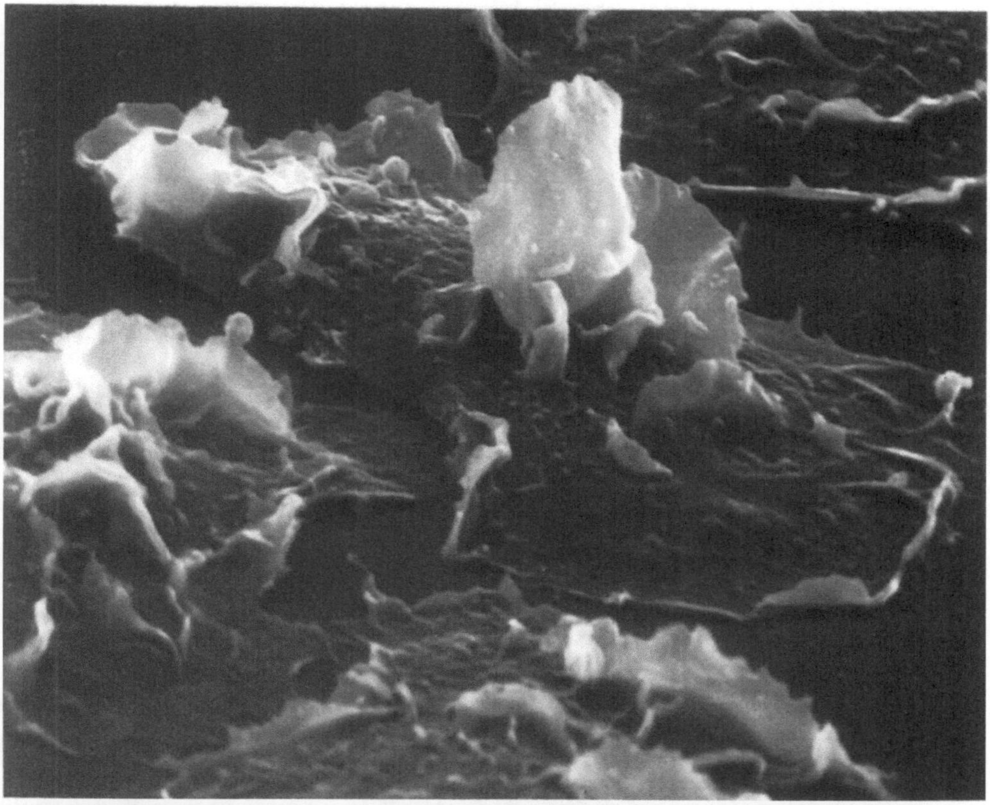

a

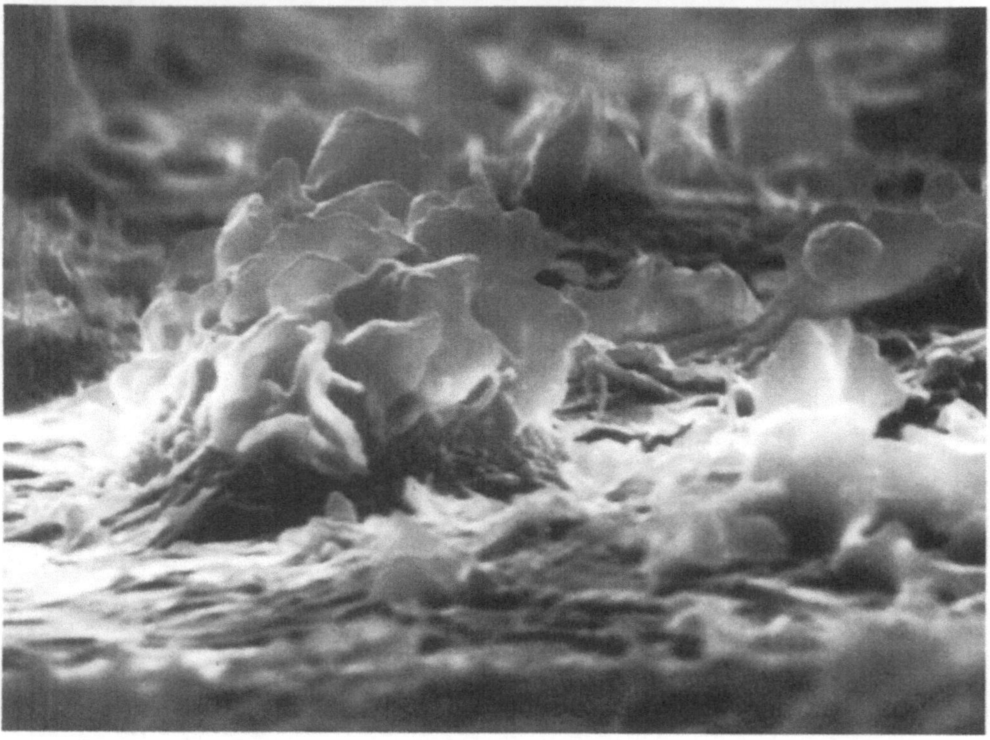

b

4. Acute Lymphoid Leukemias (ALL)

This acute lymphatic leukemia is predominantly composed of small blast cells. The motility of the cells is restricted to a moderate surface activity. The spheric cells are termed lymphoblasts because of the predominance of this cell type in the peripheral blood of this patient.

Figure 74. Leukemic population

The uniformity in cell shape is documented by both SEM (a) and cinematography (b). The surface architecture of the different cells consists not only of shorter and longer microvilli, but also of small folds and blebs. The small size of the leukemia cells is indicated by the presence of an erythrocyte (e)

a SEM: × 3000
b Phase contrast: × 2800

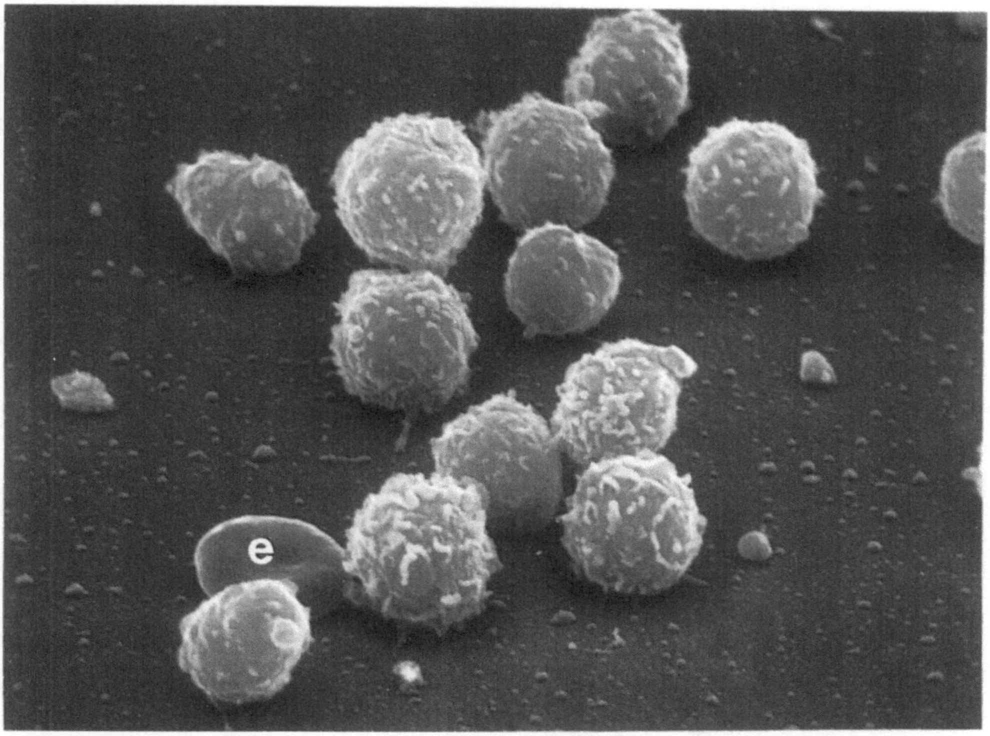

a

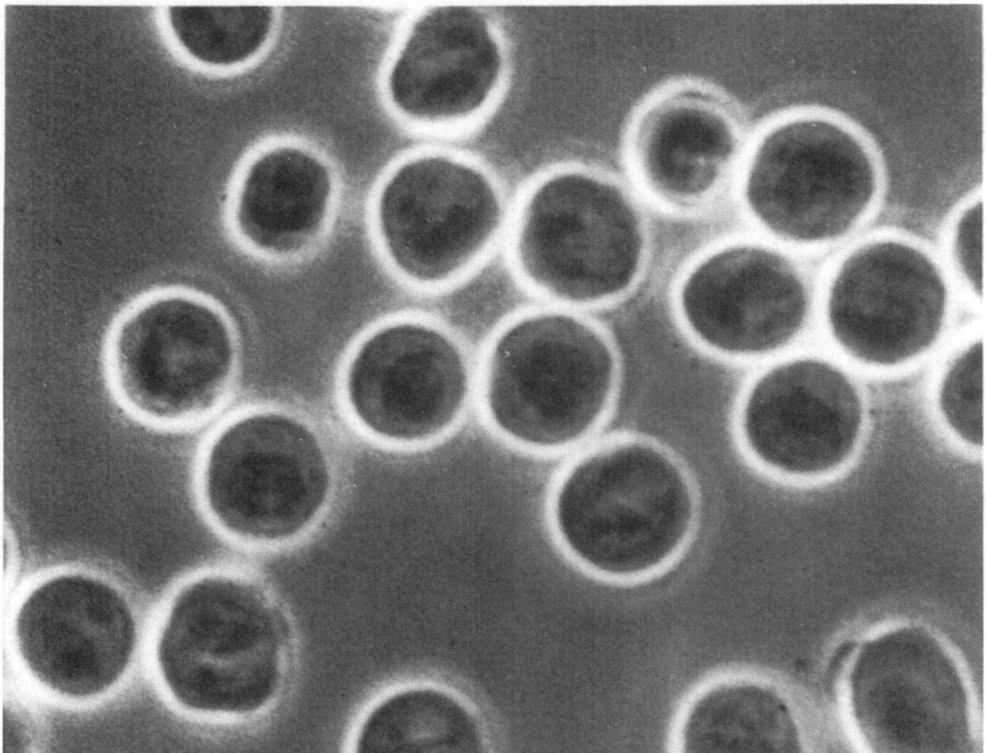
b

Figure 75. Leukemic lymphoblasts

The diversity in surface projections for the different cells is evident (a). Some slightly broader cytoplasmic projections could be the result of surface activity (a, b)

a SEM: × 8300
b SEM: × 16,000

a

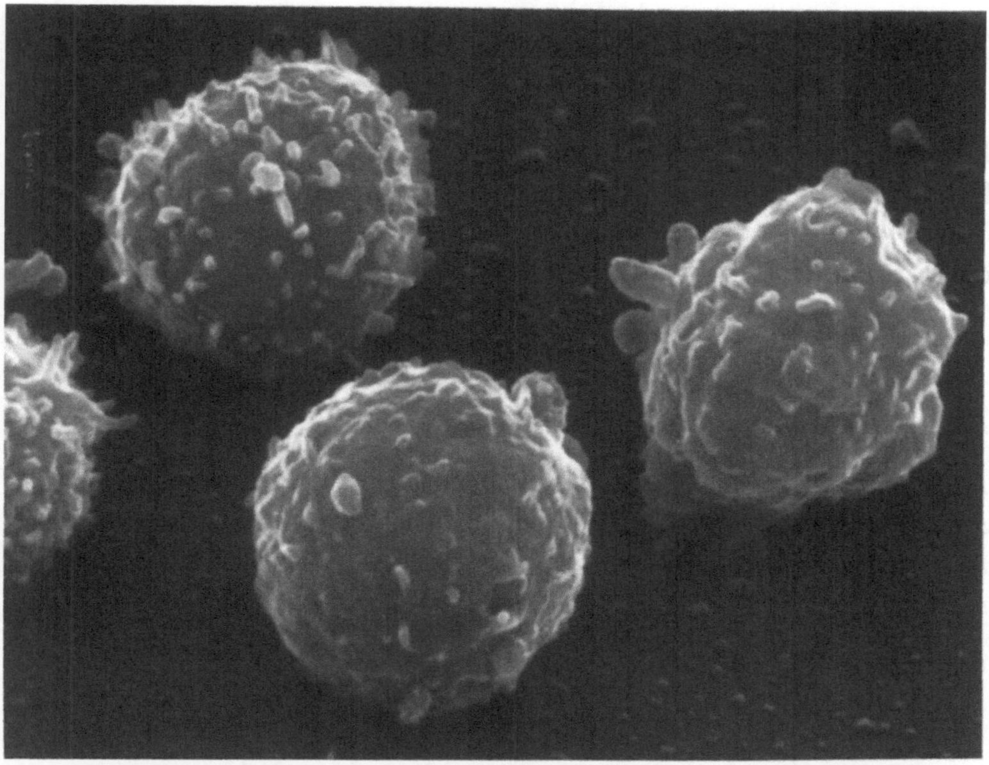

b

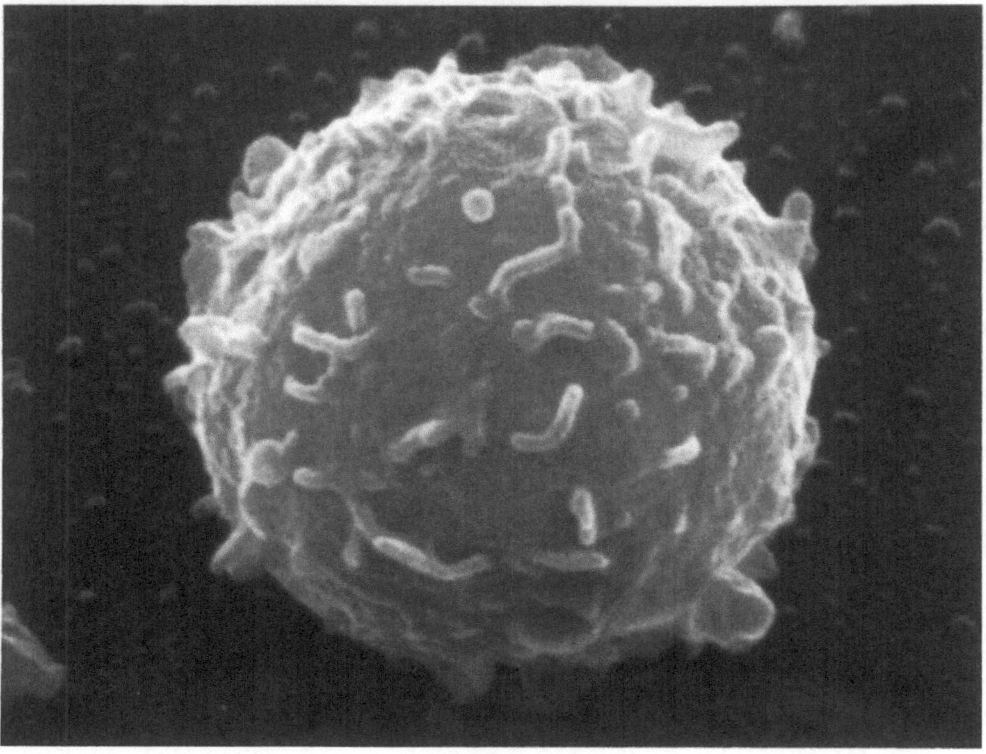

Lymphoblasts are again the main representatives of this leukemic population. The majority of the cells demonstrate surface motility, while on-spot motility in the polarized configuration and slow locomotion occur infrequently.

Figure 76. Leukemic population

Phase contrast from a time-lapse sequence verifies the predominance of spheric cells. Some have an irregular periphery, indicative of surface activity. According to cinematographic evidence, the one polarized cell is locomotive

Phase contrast: $\times$ *3400*

Figure 77. Leukemic lymphoblasts *(see page 130)*

The surface morphology of the spheric cells which, according to cytologic data, can be considered to be lymphoblasts, is different for indicidual cells: some show finger- and stub-like microvilli, some more small folds and a few blebs. This, again, demonstrates the diversity of surface extensions on cells within the same population

a SEM: $\times$ *7300*
b SEM: $\times$ *9100*

Figure 78. Leukemic lymphoblasts *(see page 131)*

The divergency in surface extensions of different cells is presented in higher magnifications

a SEM: $\times$ *27,900*
b SEM: $\times$ *28,700*

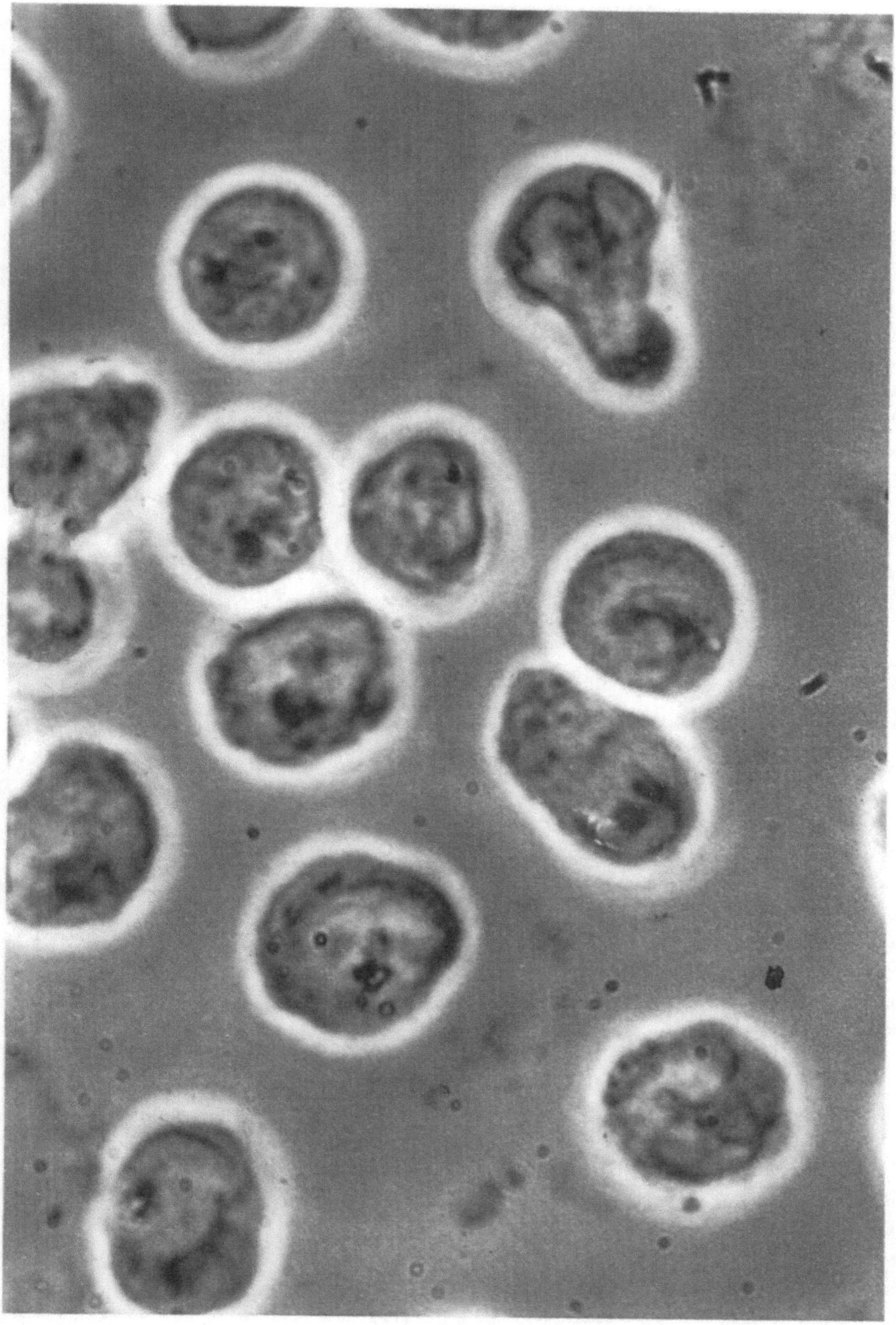

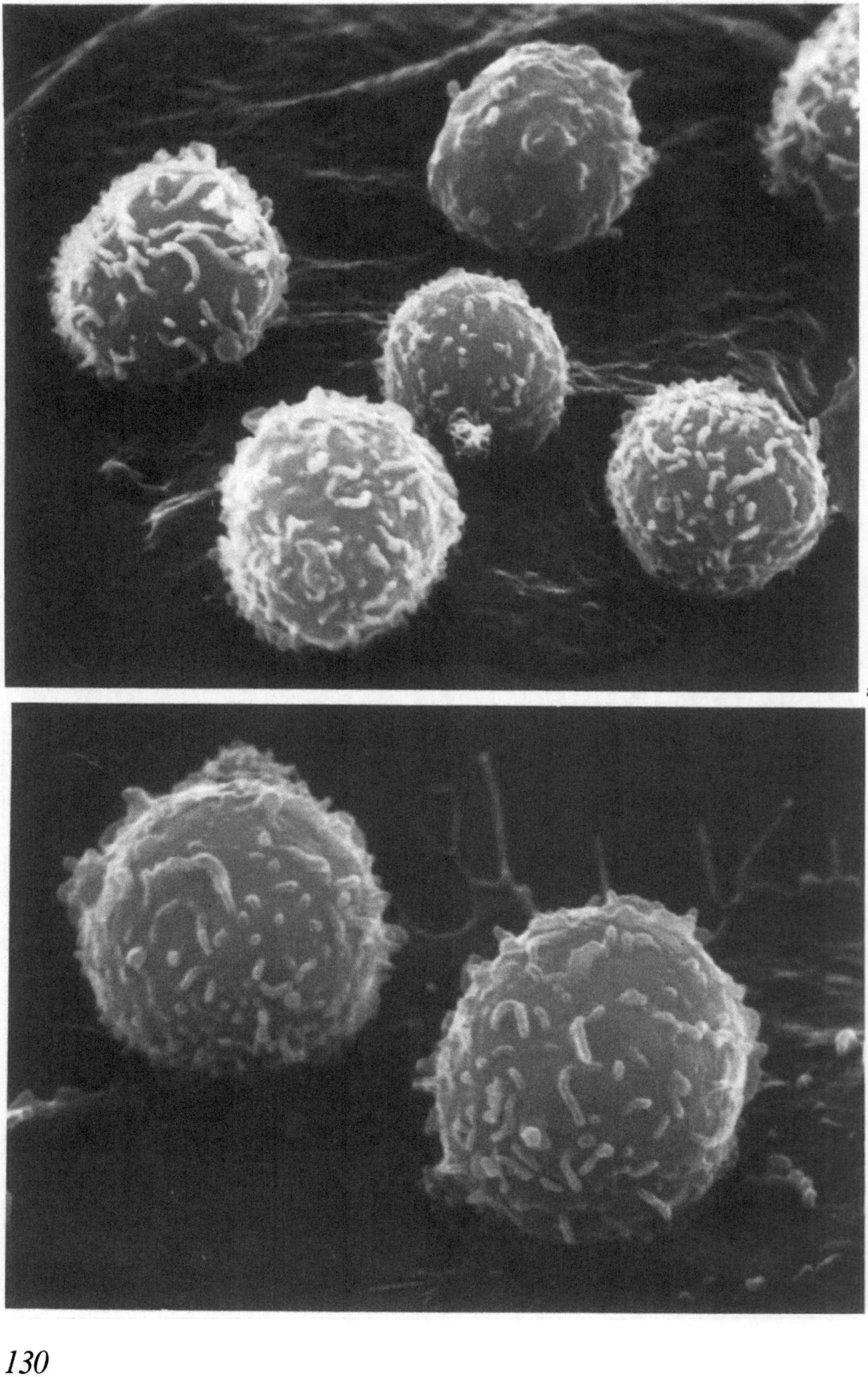

a

b

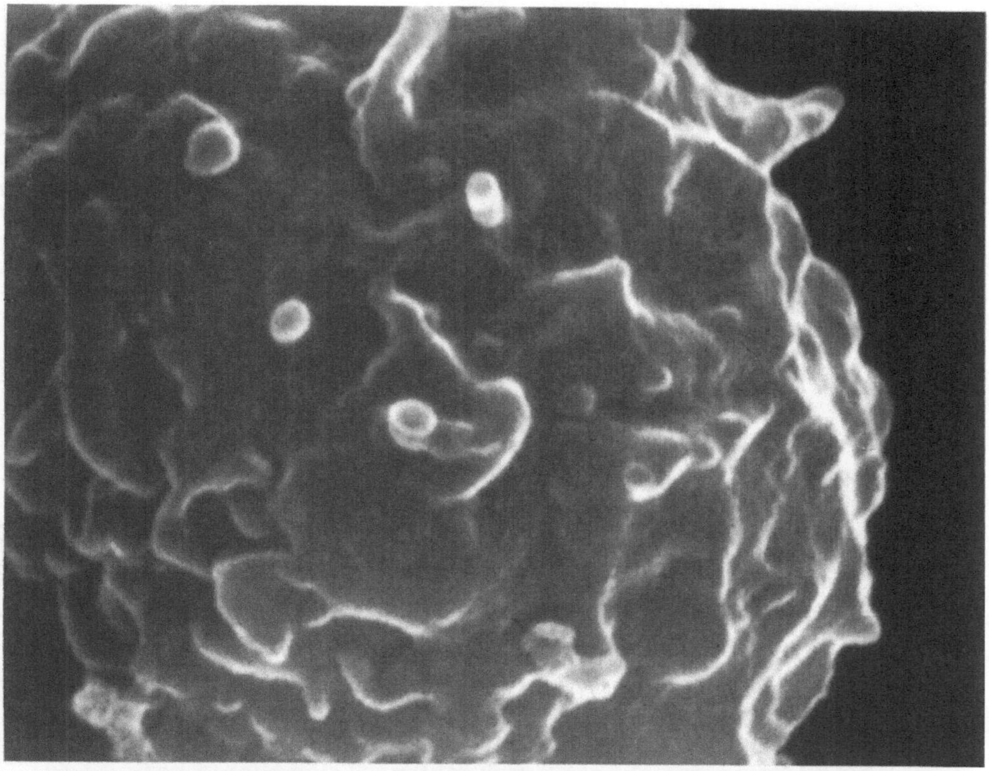

a

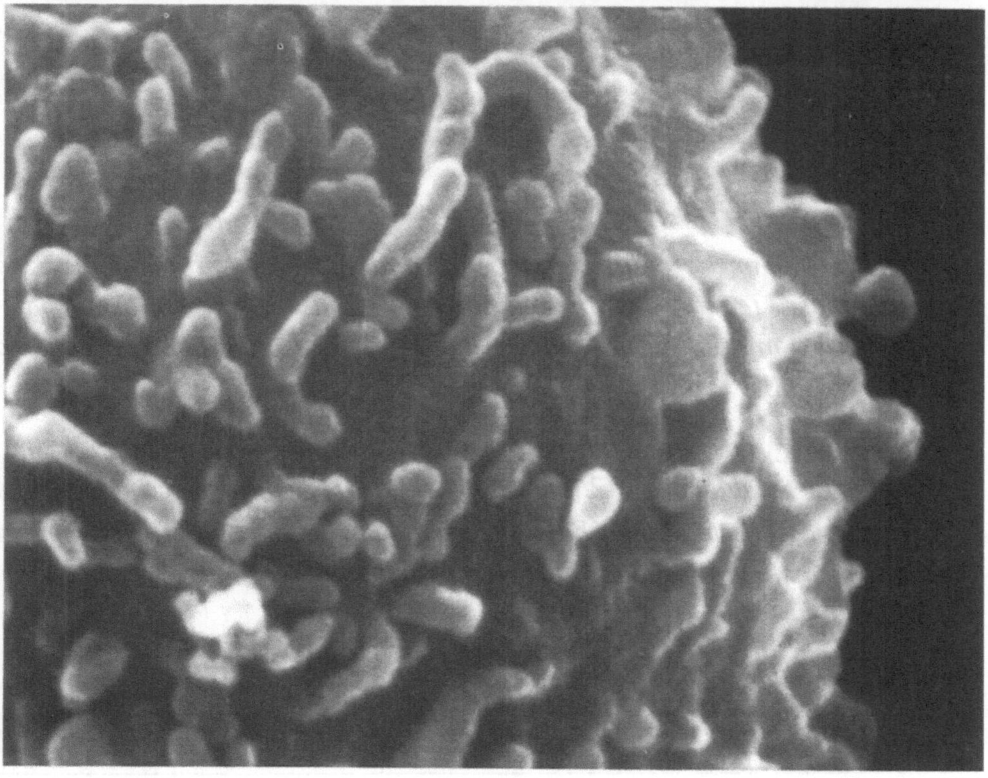

b

Figure 79. Leukemic lymphoblasts

a A spheric and a polarized lymphoblast are attached to a spread monocyte. The polarized configuration suggests on-spot motility

b Another polarized blast cell adhering with its foot to the edge of a spread monocyte and photographed at a tilt angle of 85°

a SEM: × *11,000*
b SEM: × *13,600*

Figure 80. Leukemic lymphoblasts *(see page 134)*

Another example for on-spot activity in the polarized form (a), where a spread cell is again utilized for attachment instead of the surrounding free glass surface. The presence of cytoplasmic contacts at the front part (arrows) could be indicative of a change from on-spot motility to locomotion. The especially round and stubby posterior part is covered with short microvilli, also visible in the TE micrograph (b)

a SEM: × *17,300*
b TEM: × *12,400*

Figure 81. Leukemic lymphoblasts *(see page 135)*

a Several attachment points on the tail and the body of this polarized lymphoblast, together with medium-sized folds at the anterior part, indicate locomotion. The short, round tail is again densely covered with microvilli

b The drawing represents the pathway of a locomotive lymphoblast that covered the distance of 42 µm in 38.4 min at a speed of ca 1 µm/min. The cells of this leukemia have the particularity of temporarily moving sideways instead of ahead. During this unorthodox behavior, the polarization of the cells is maintained

a SEM: × *15,000*
b Drawing from a sequence on 16-mm film

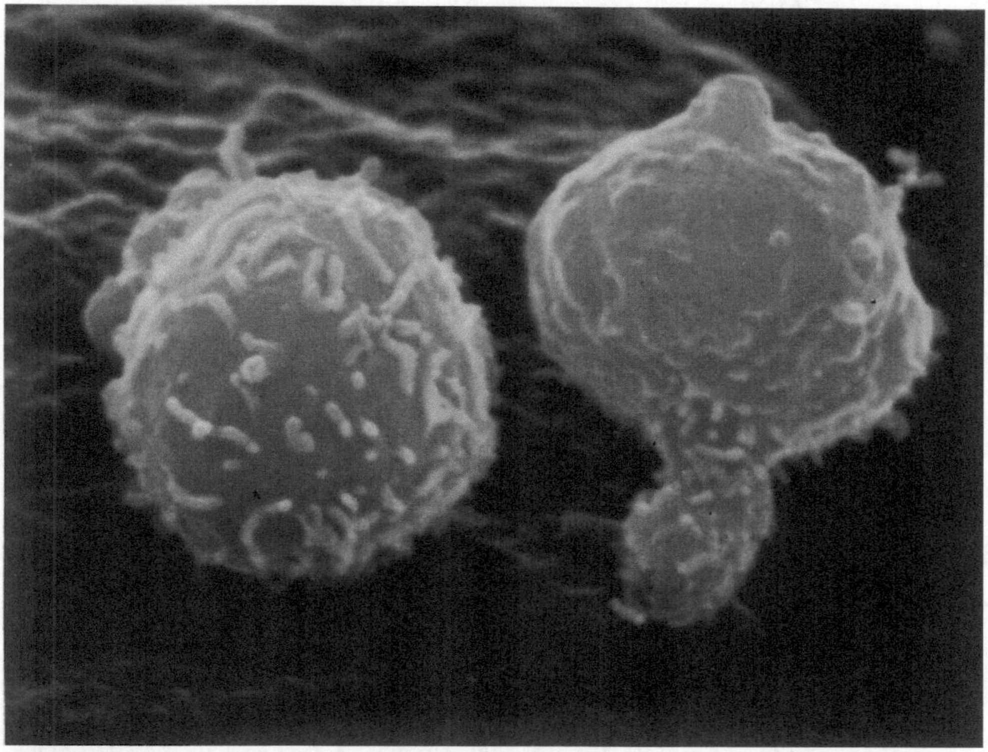

a

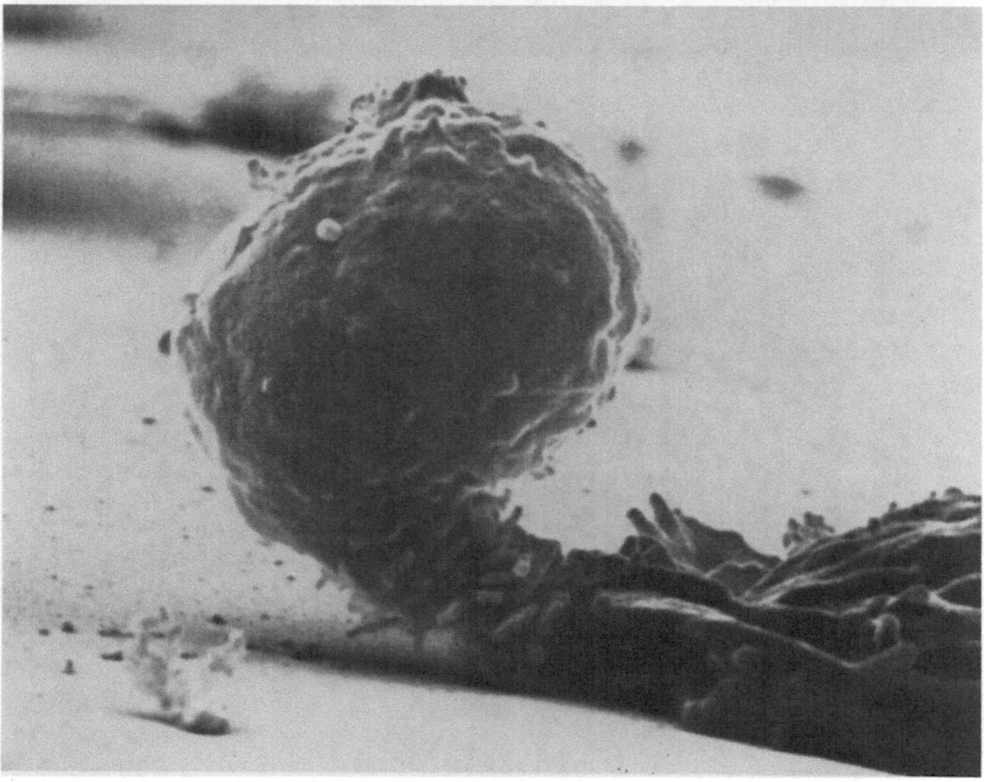

b

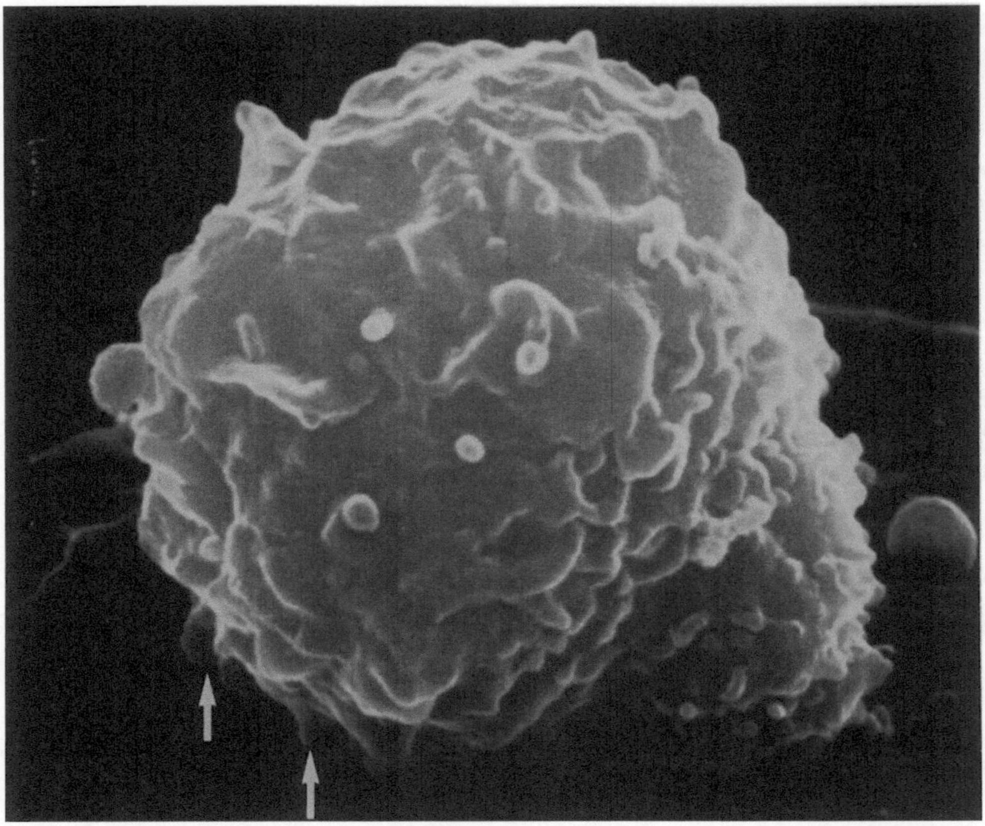

a

b

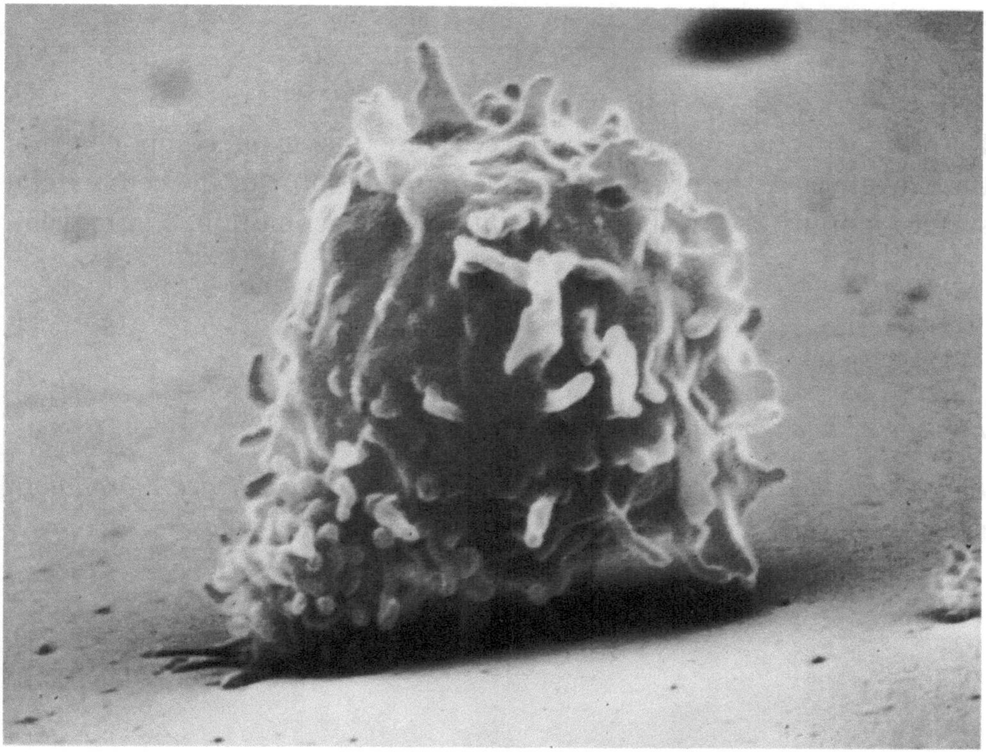

a

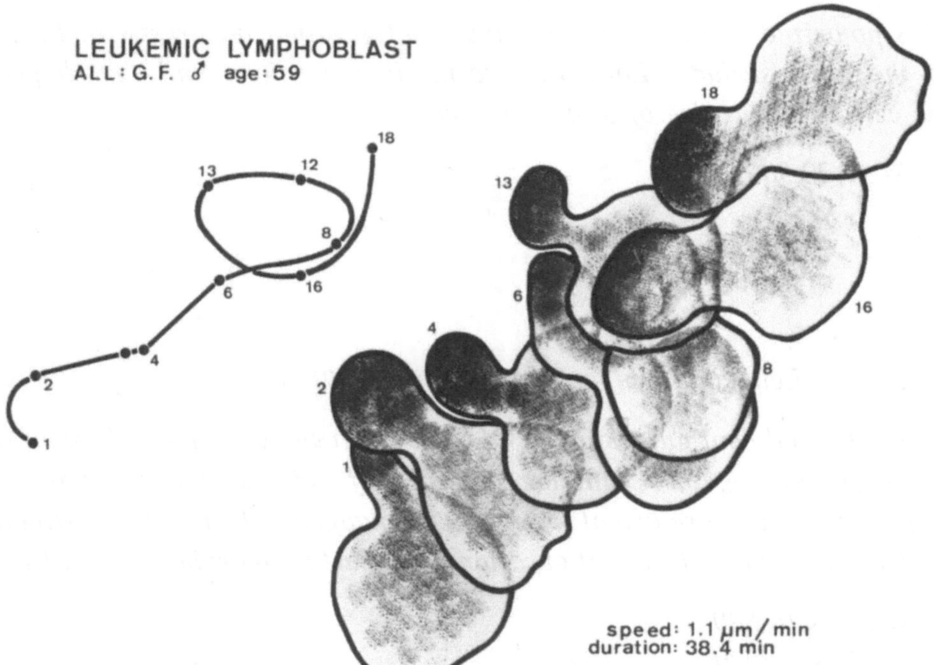
LEUKEMIC LYMPHOBLAST
ALL: G. F. ♂ age: 59
18
13
12
8
6
16
4
2
1
18
13
6
4
2
1
16
8
speed: 1.1 μm/min
duration: 38.4 min

b

Rather small and mostly spheric cells are the main components of this leukemia. As recorded by cinematography most of the cells exhibit surface motility, a few on-spot motility, while locomotion is infrequent.

Figure 82. Leukemic population

Individual spheric cells again show a rather irregular surface morphology with shorter and longer microvilli, ridges, and smaller folds

SEM: × 9600

Figure 83. Leukemic lymphoblast *(see page 138)*

This polarized cell fixed in the blast configuration adheres with its foot to the glass surface. Long cytoplasmatic extensions and the shape of the cell are indicative of on-spot activity

SEM: × 22,300

Figure 84. Leukemic lymphoblasts *(see page 139)*

The polarized configuration, the close contact between the cell body and the substrate together with the presence of veils at the anterior part, suggest that these are two of the few cells engaged in locomotion. Although both cells have the blast configuration, their surface morphology is different

a SEM: × 16,800
b SEM: × 14,000

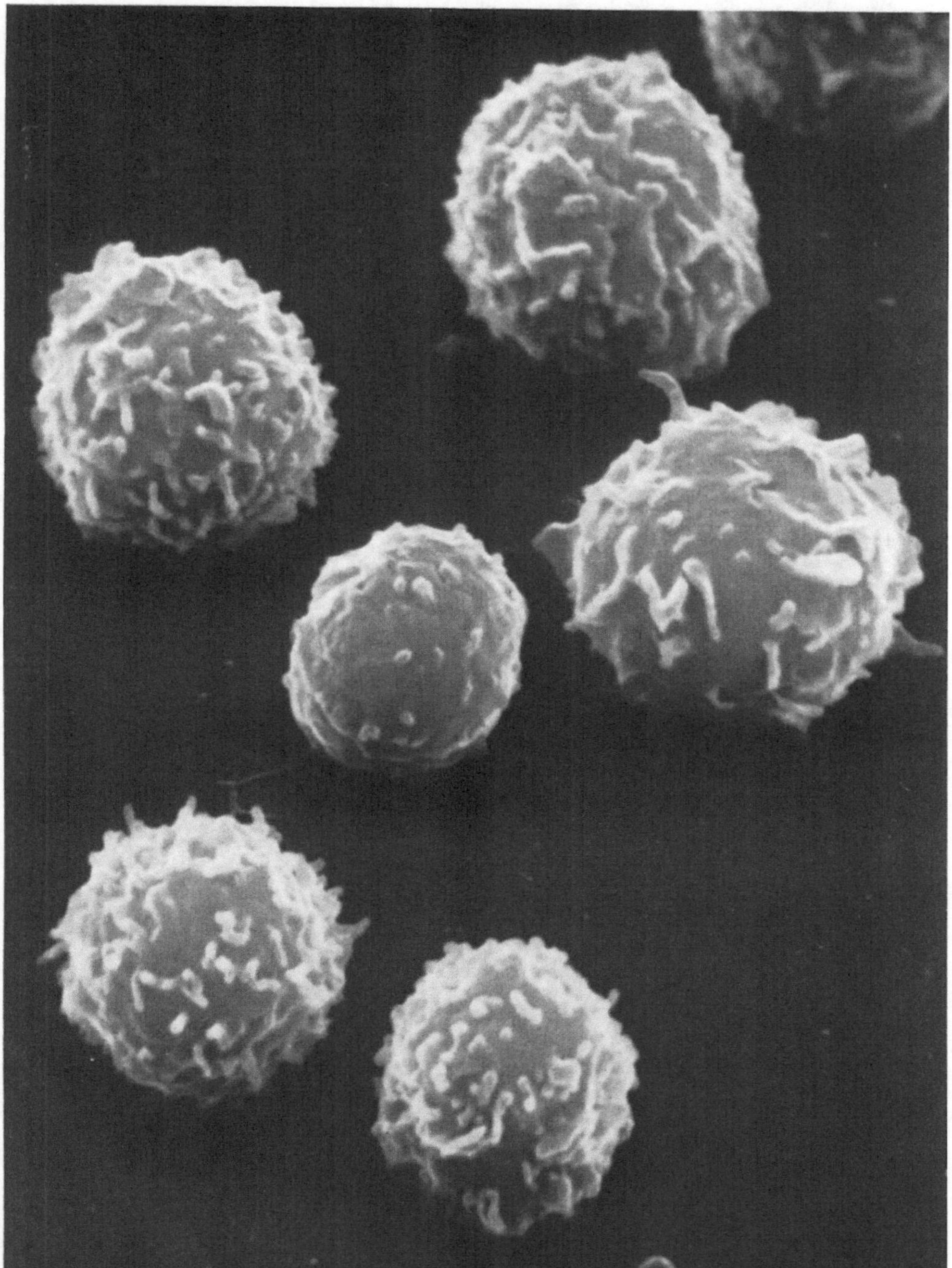

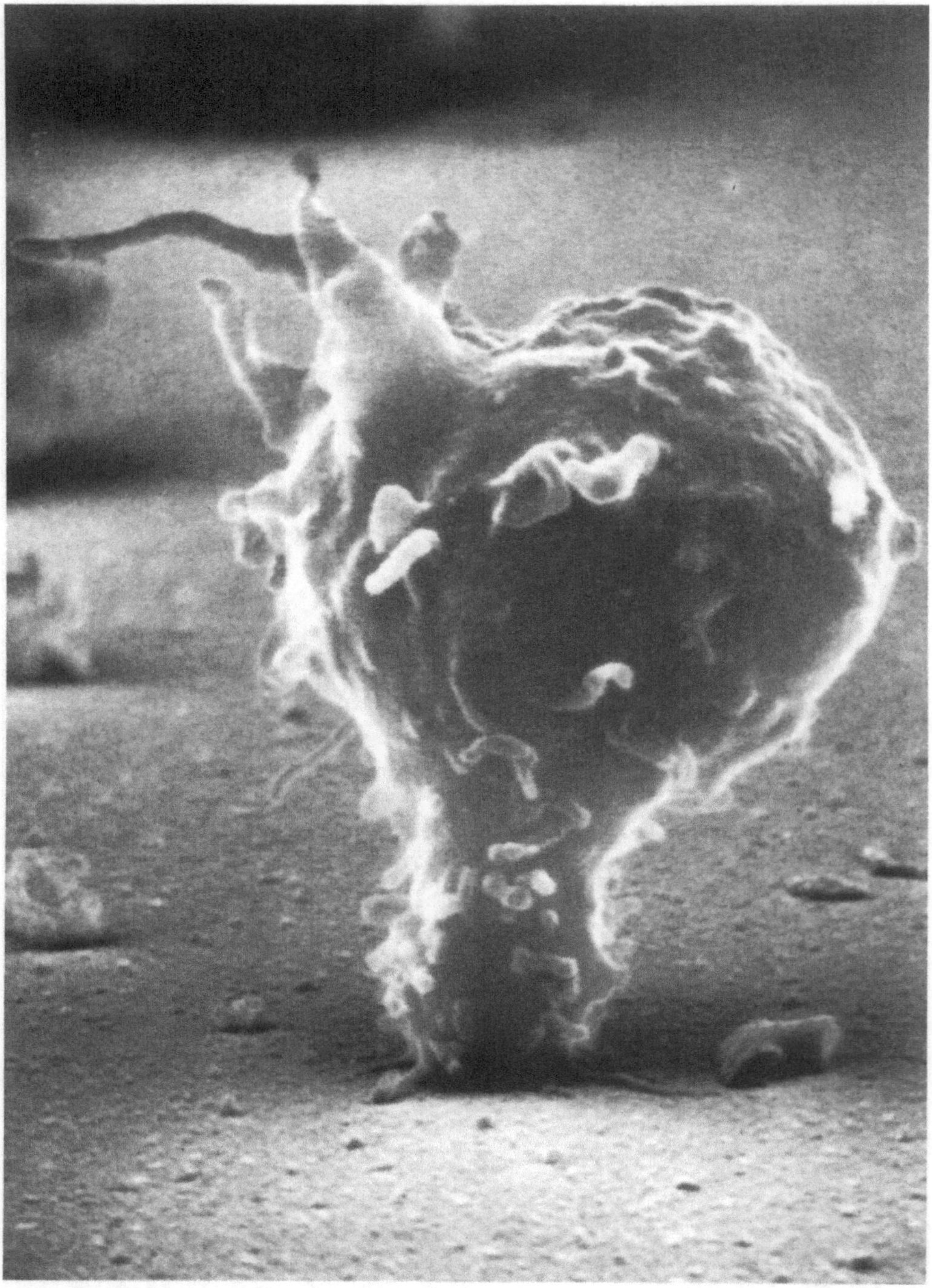

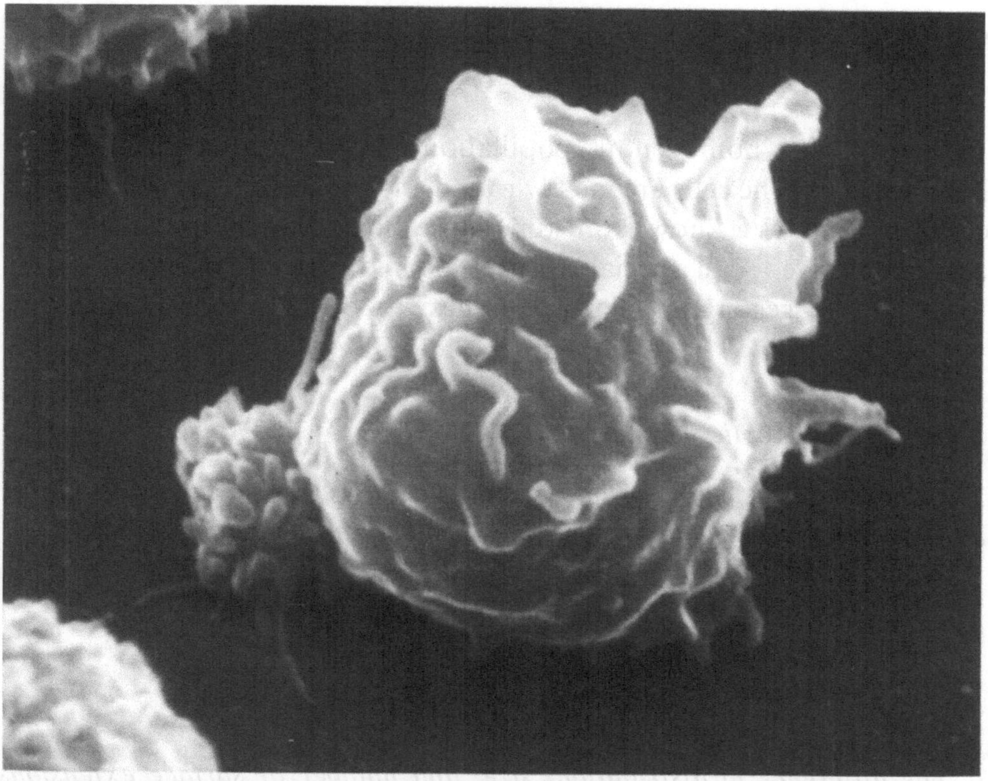

a

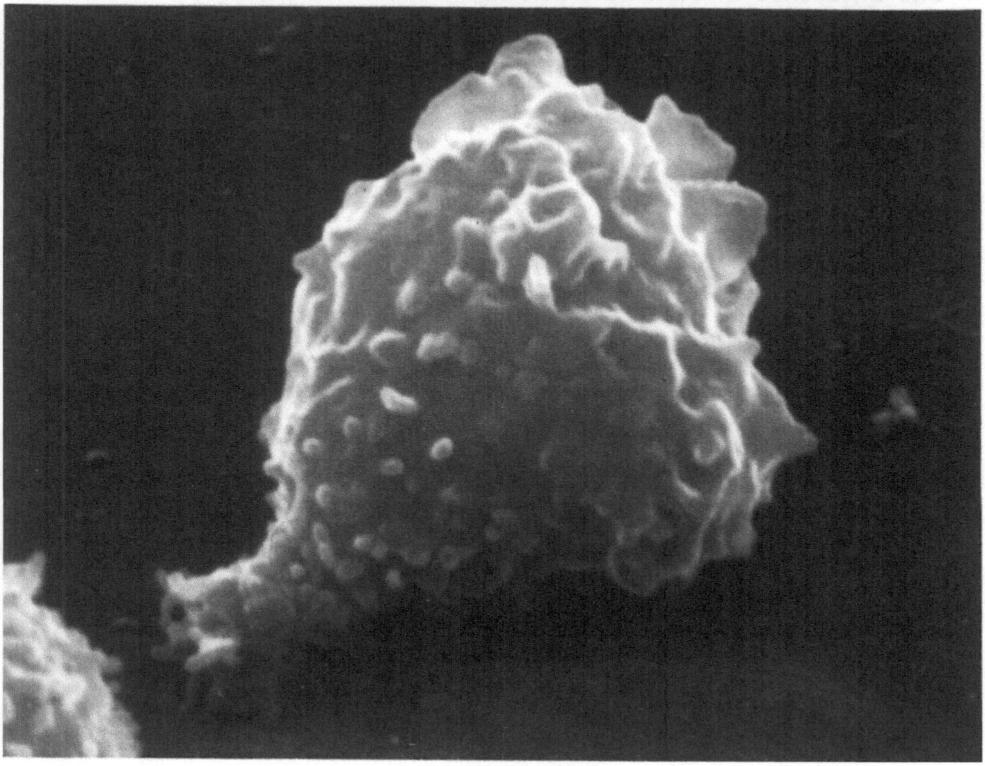

b

This again is a homogenous population consisting almost entirely of lymphoblasts. The cells are mostly round and only a few have a polarized configuration. We know from time-lapse films that the polarized cells are locomotive. As a special feature, locomotive cells often have very long tails, suggesting an increased adhesivity of the posterior part.

Figure 85. Leukemic population

The phase-contrast picture taken from time-lapse sequences confirms the uniformity of the population that includes only a few polarized cells

Phase contrast: × 1700

Figure 86. Leukemic population *(see page 142)*

a The difference in distribution, size, and number of surface extensions of individual cells is evident
b The arrow points to the cell that shows the beginning of a polarized shape

a SEM: × 7200
b TEM: × 4600

Figure 87. Leukemic lymphoblasts *(see page 143)*

The SE micrograph (a) shows a lymphoblast with a drawn-out tail that is very similar to the one pictured in the phase-contrast photograph from a time-lapse sequence (b). The cell in the foreground of the SEM picture is a thrombocyte (th). The spheric cell cannot be identified by SEM but could be a lymphoblast because of the predominance of this cell type

a SEM: × 5700
b Phase contrast: × 3800

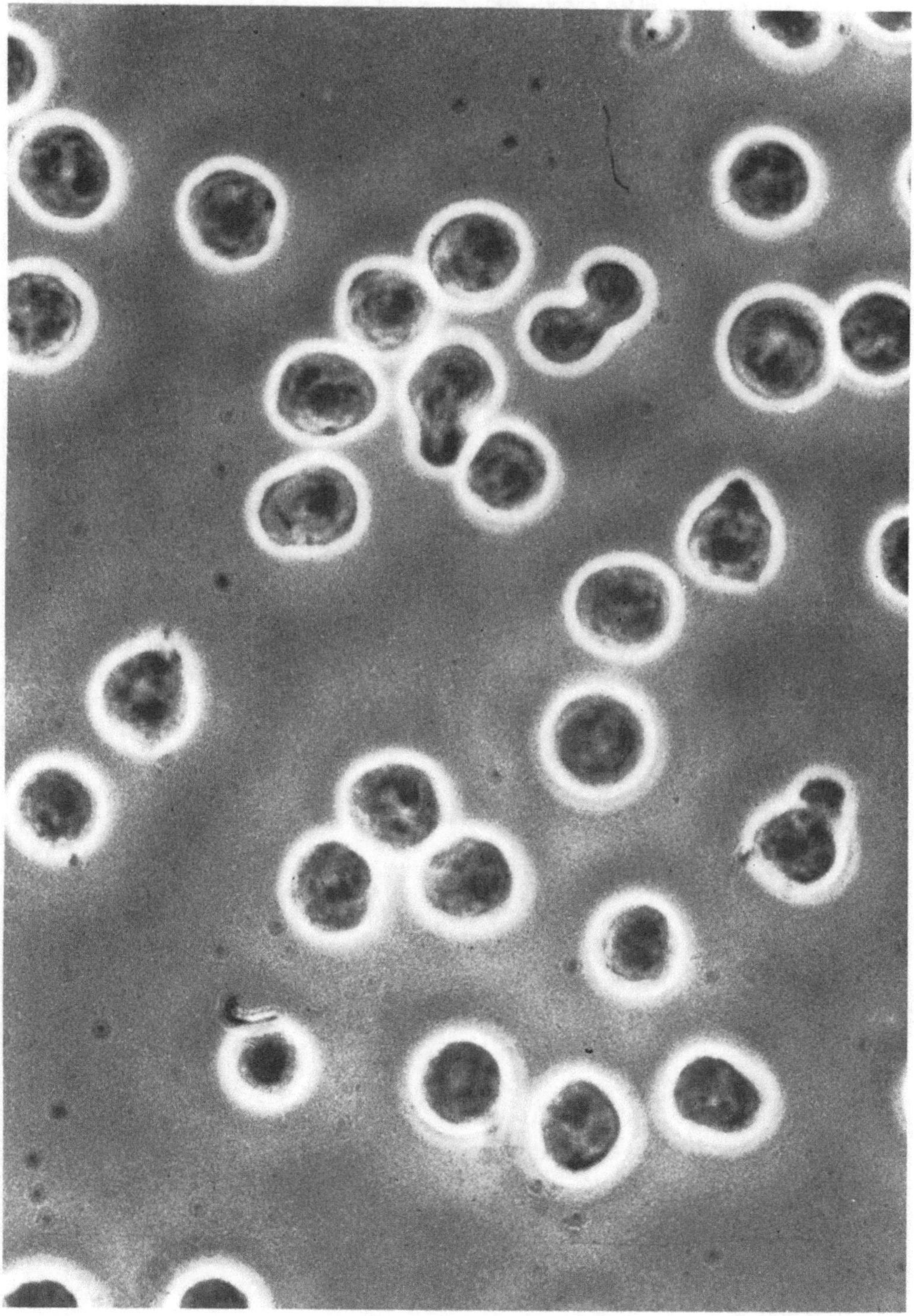

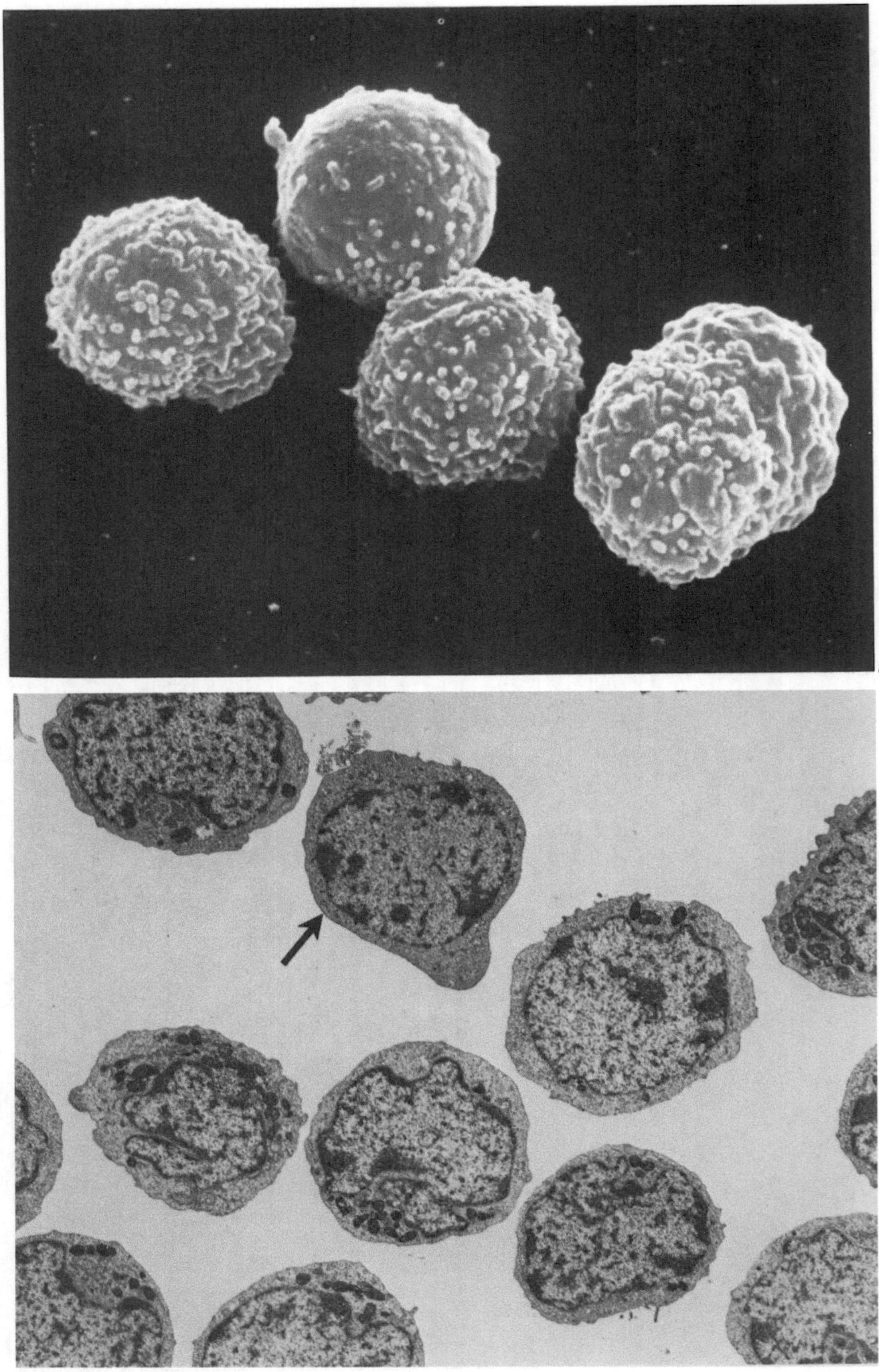

a

b

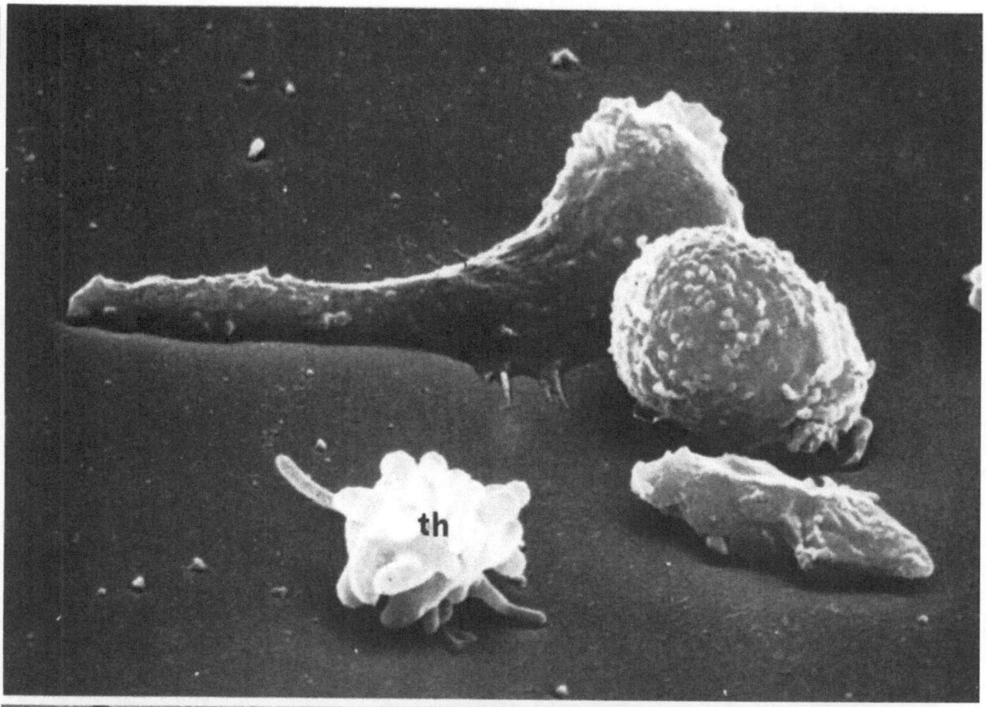

a

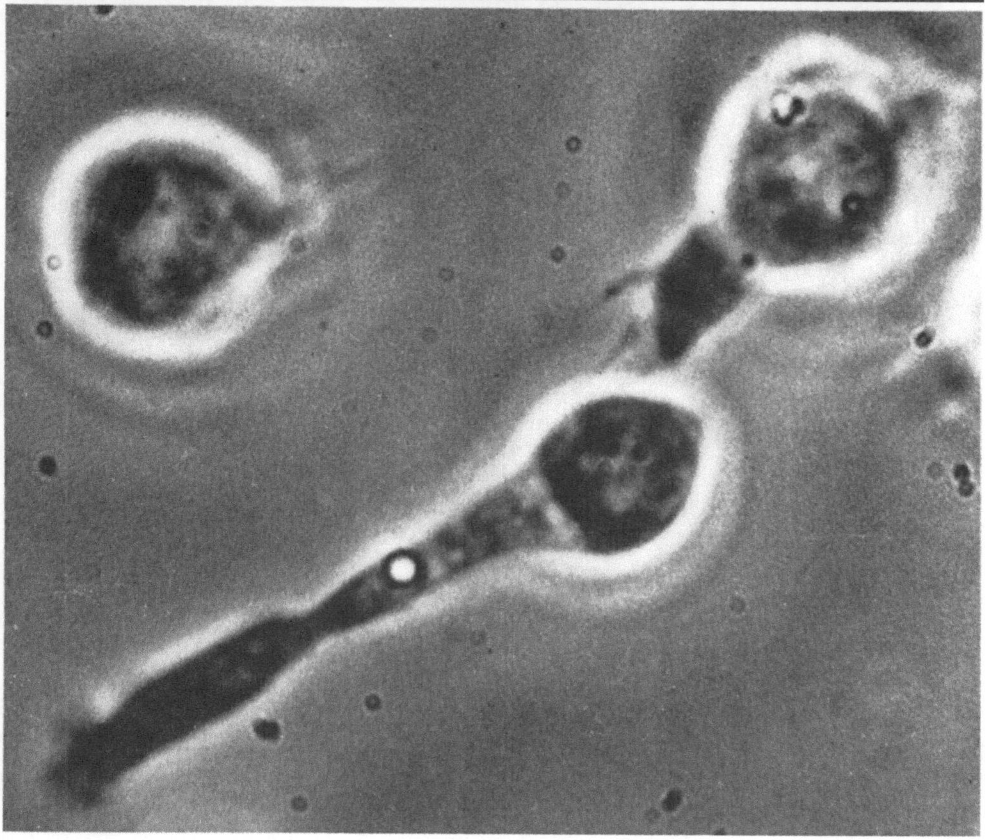

b

5. Chronic Lymphoid Leukemia (CLL)

This chronic lymphatic leukemia was diagnosed several years ago and the patient is being treated intermittently with prednisone. The population predominantly consists of smaller and larger lymphocytes that mainly demonstrate surface motility.

Figure 88. Leukemic population

Again a population where the variation in surface architecture of different cells is evident. In this photograph the cells, with one exception, are adhering to a spread monocyte which has retained its centrally placed large folds

SEM: × 8900

Figure 89. Leukemic population *(see page 148)*

Larger and smaller leukemia cells are presented in this phase-contrast photograph taken from a time-lapse sequence

Phase contrast: × 4400

Figure 90. Leukemic lymphocytes *(see page 149)*

Among spheric cells there is one with a polarized configuration. It has made contact with a spheric leukemia cell by means of small cytoplasmic extensions at the front

SEM: × 10,900

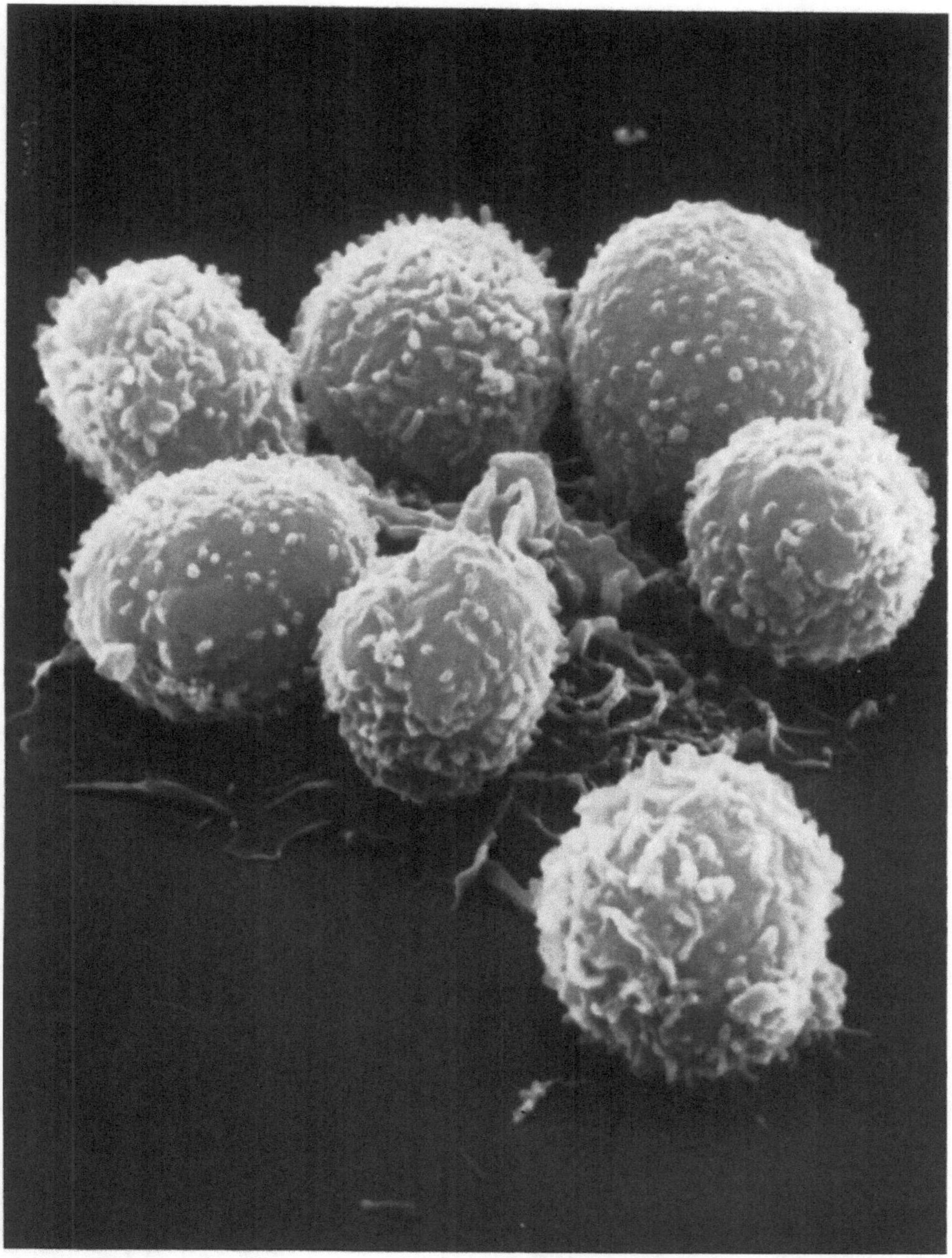

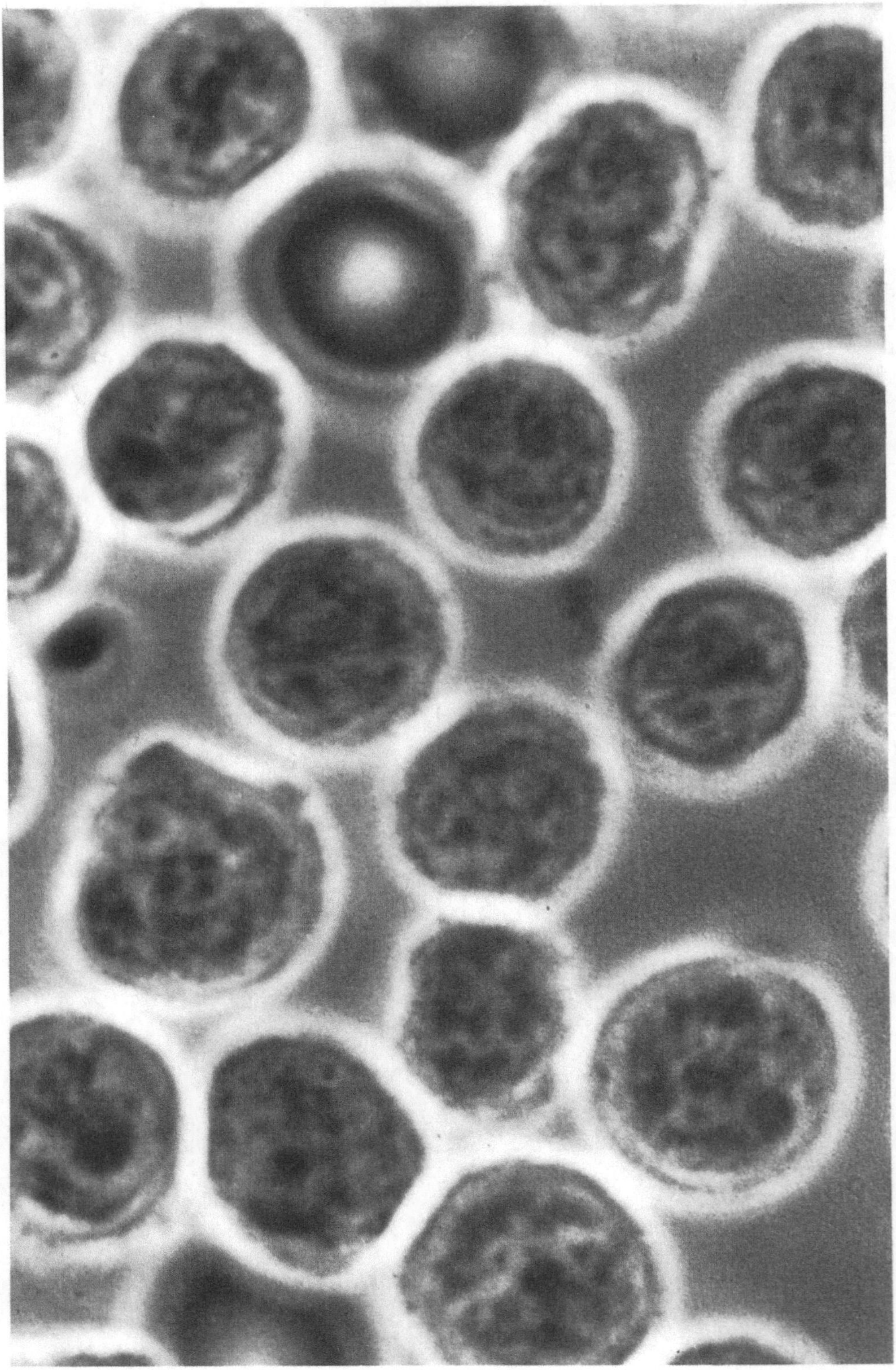

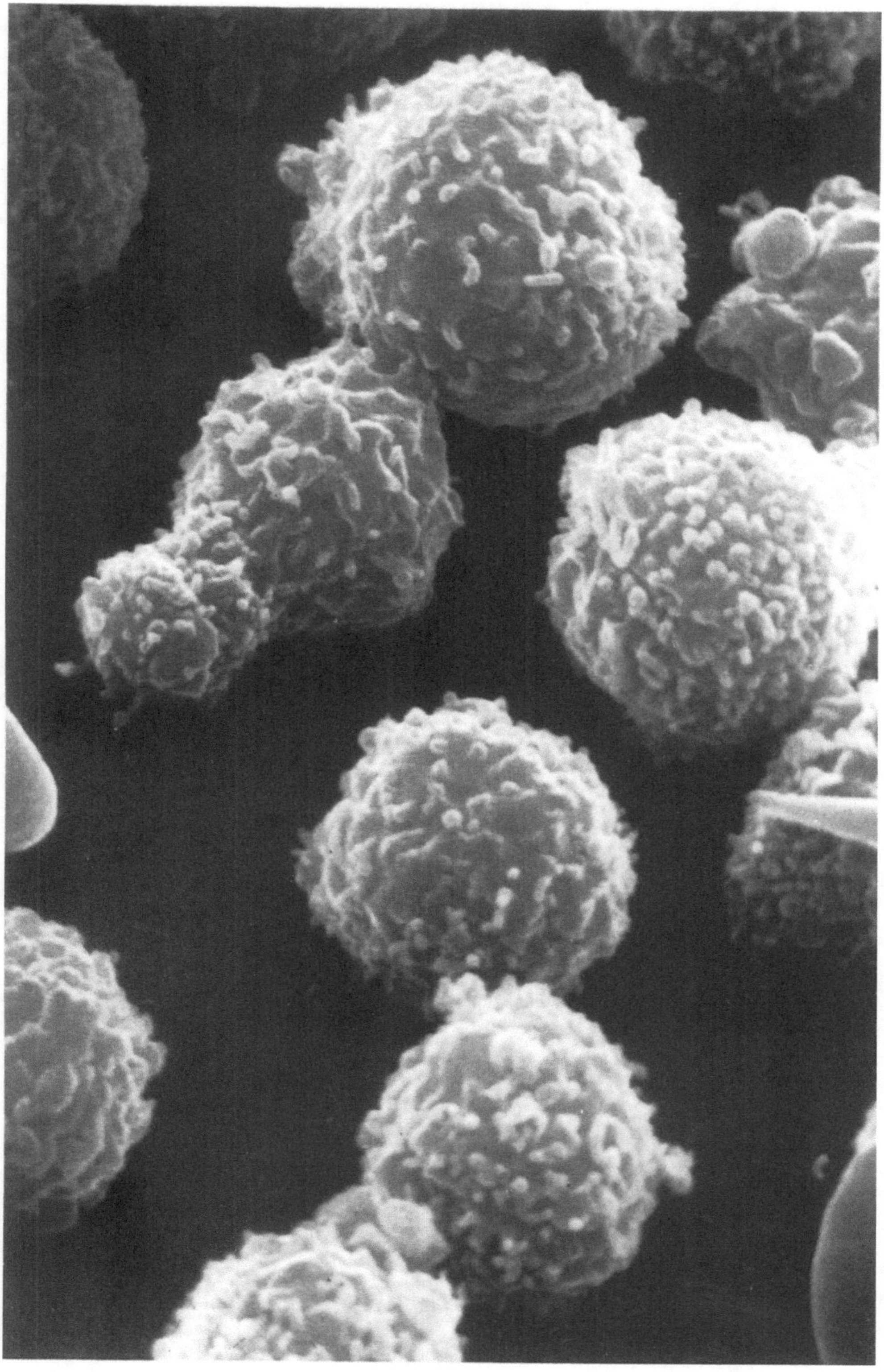

6. Unclassifiable Leukemias

This leukemia is characterized by a very heterogenous population: large blast cells, lymphocytes and granulocytes are the main constituents, but promyelocytes, myelocytes, and monocytes are also recognizable. The leukemia has been termed "unclassifiable," as neither cytology and cytochemistry nor morphology at the TEM and SEM level allowed certain identification of the blast cells as being of either myeloid or lymphoid origin. Microcinematography was of no help either in attempts at classification, as blast cells, irrespective of their origin, locomote in a polarized shape in the blast pattern of locomotion.

Figure 91. Leukemic population

In this phase-contrast photograph from a time-lapse film, some of the representatives of the heterogenous population are present: large blast cells (bl) and lymphocytes (ly) both in polarized and spheric shape, and a promyelocyte (pm)

Phase contrast: × 2400

Figure 92. Leukemic population *(see page 154)*

The heterogeneity of the population is expressed in the difference of size, shape, and surface morphology

a SEM: × 6700
b SEM: × 7500

Figure 93. Leukemic blast cell *(see page 155)*

At a higher SEM magnification, a blast cell in its typical polarized shape is adherent, together with other cells, on a spread monocyte. SEM does not allow identification of polarized blasts as being of either lymphoid or myeloid origin

SEM: × 20,600

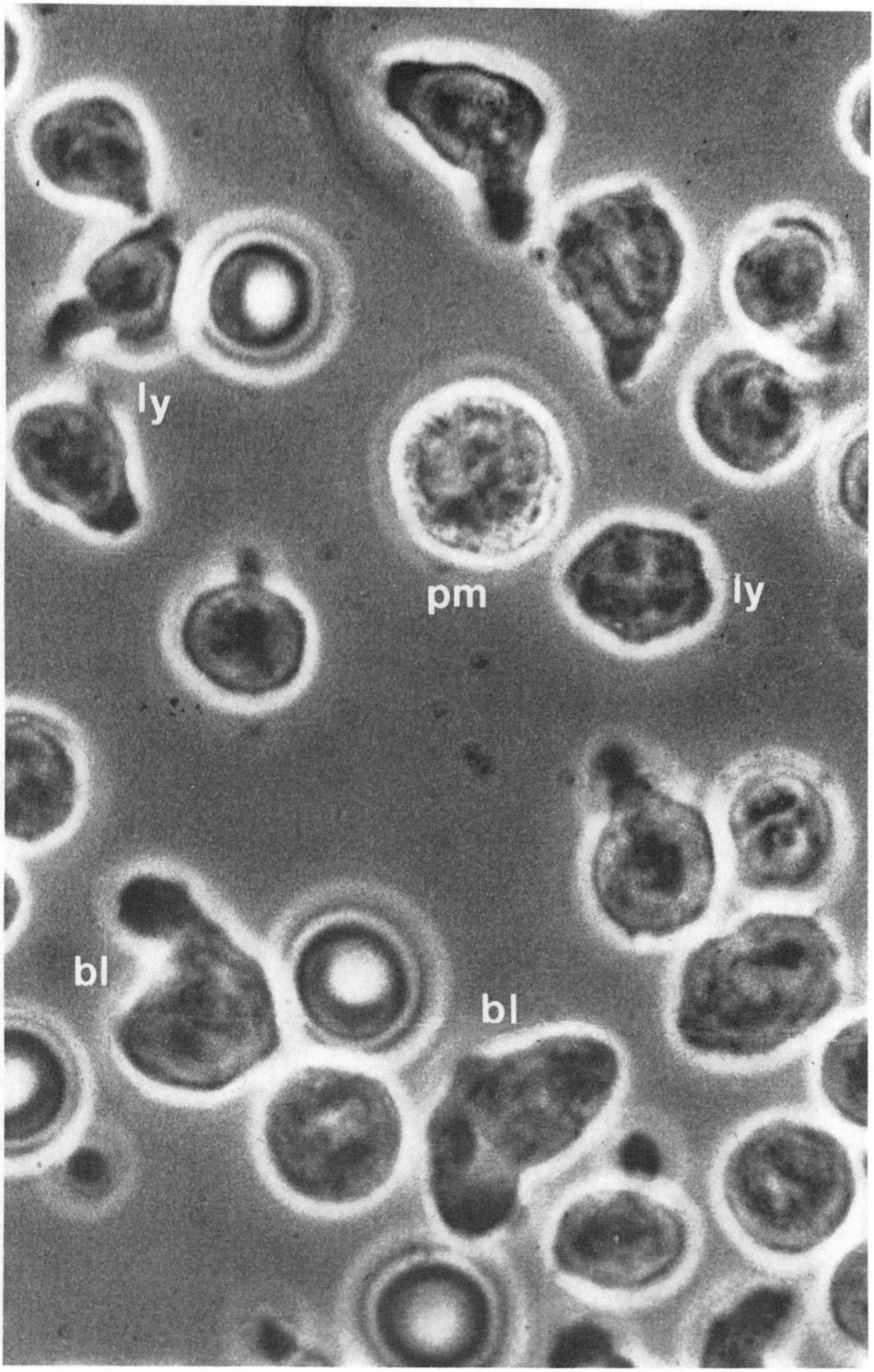
ly
pm
ly
bl
bl

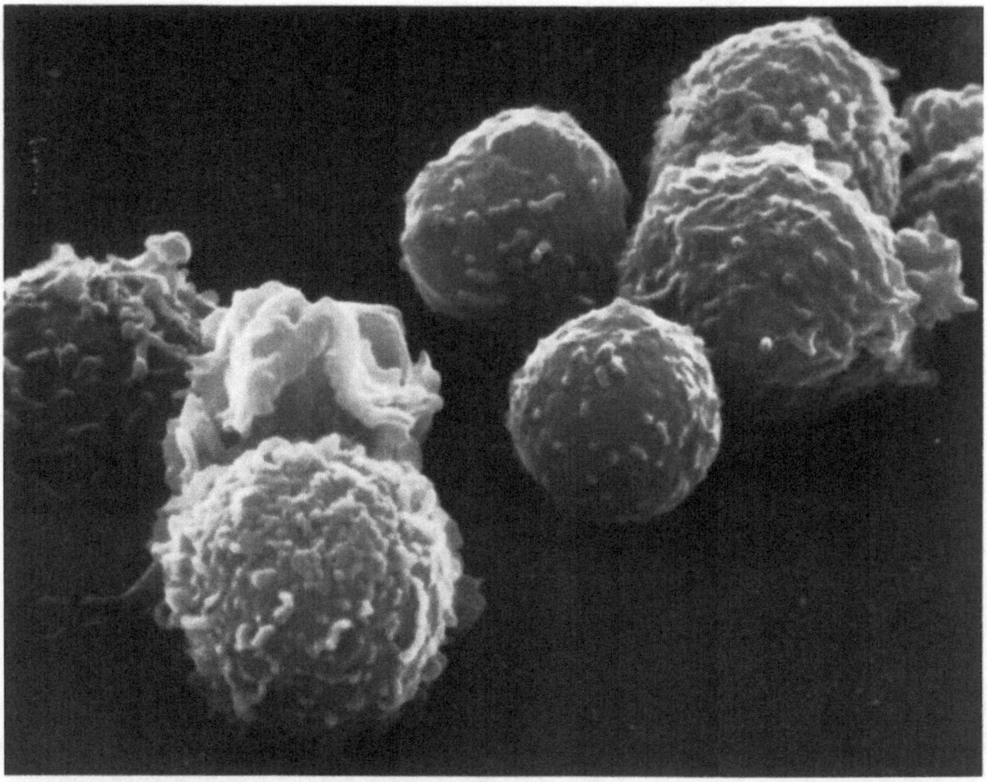

a

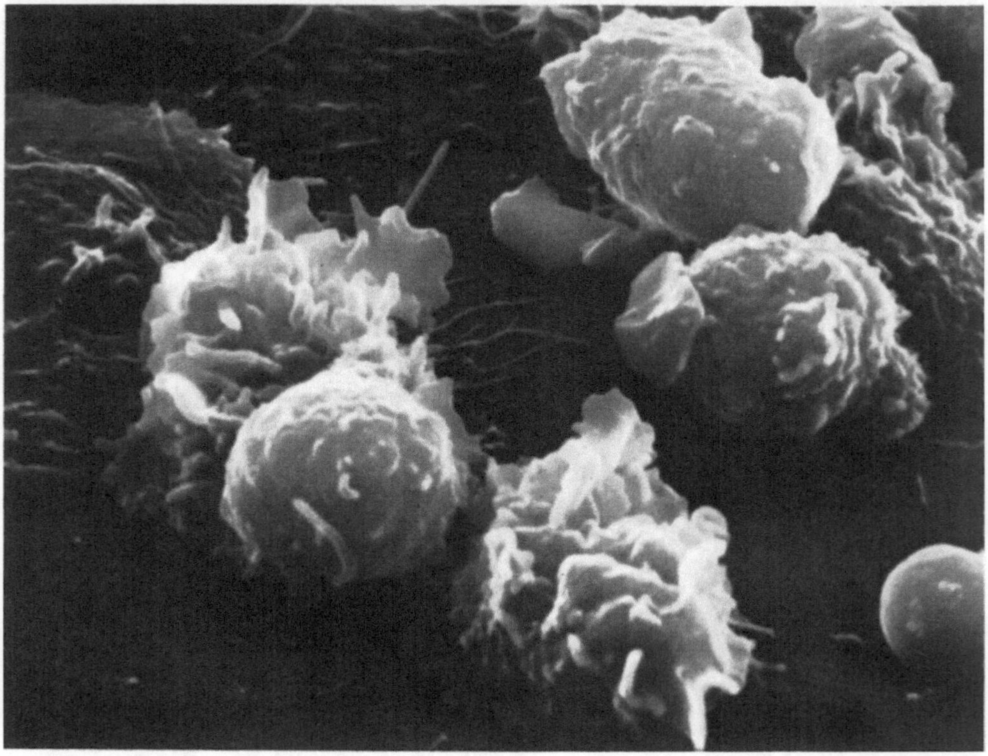

b

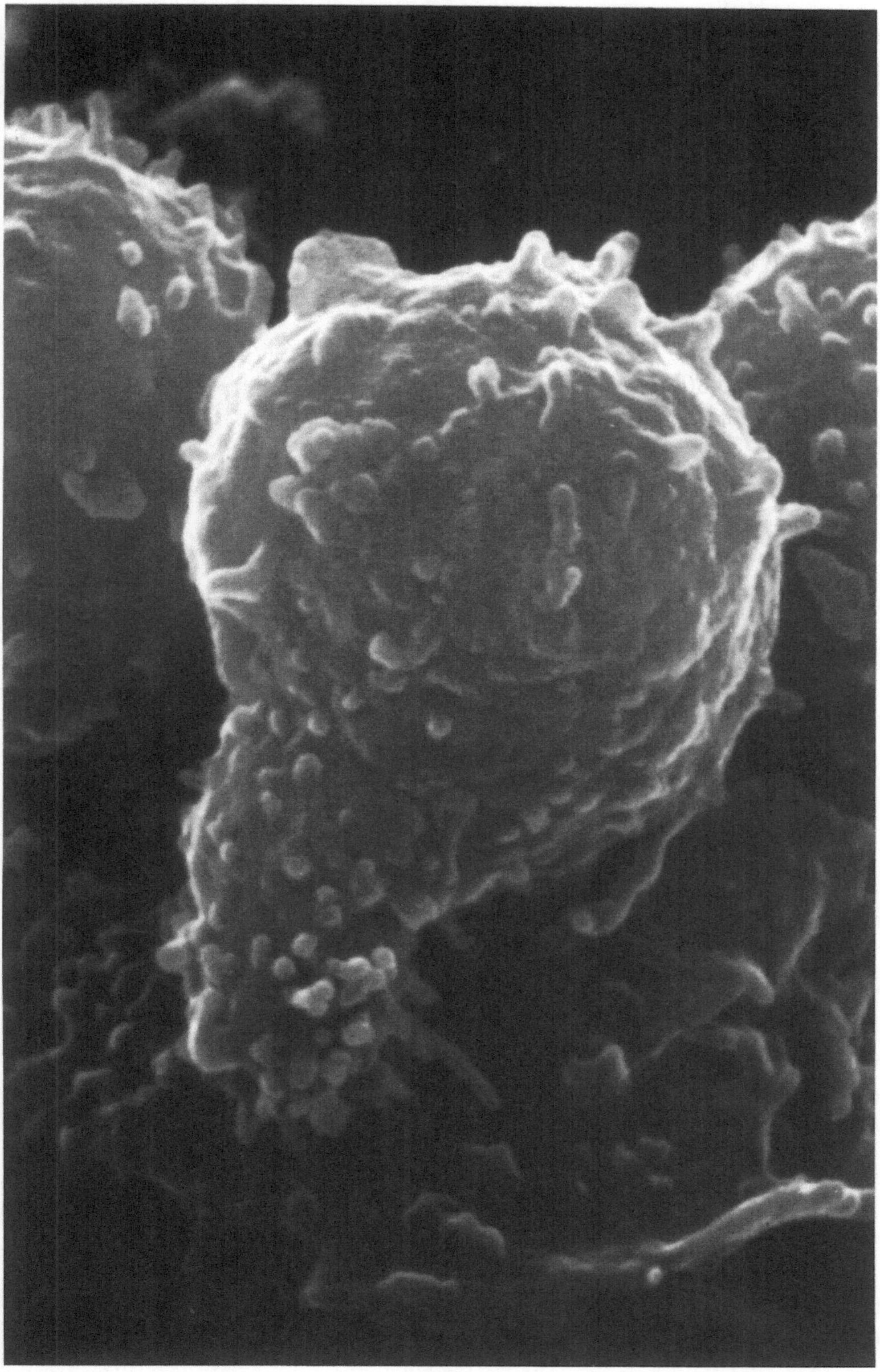

Figure 94. Unclassifiable cells

Cells of such a bizarre shape (a and b) represent a cell type not encountered in the other leukemias presented so far, but frequently seen in this population. The cells are nonpolarized and broadly adherent to the substrate, sometimes on a large, thin cytoplasmatic attachment plate (b). As the surface architecture consists of very large folds and veils, these cells could originate from the monocytoid series. However, in contrast to other representatives from the same lineage, they have not flattened, and we are therefore unable to classify them

a SEM: × 10,800
b SEM: × 14,400

Figures 95 and 96. Unclassifiable cells *(see pages 158, 159)*

Two more representatives of the type described for Figure 94

95 SEM: × 20,600
96 SEM: × 20,200

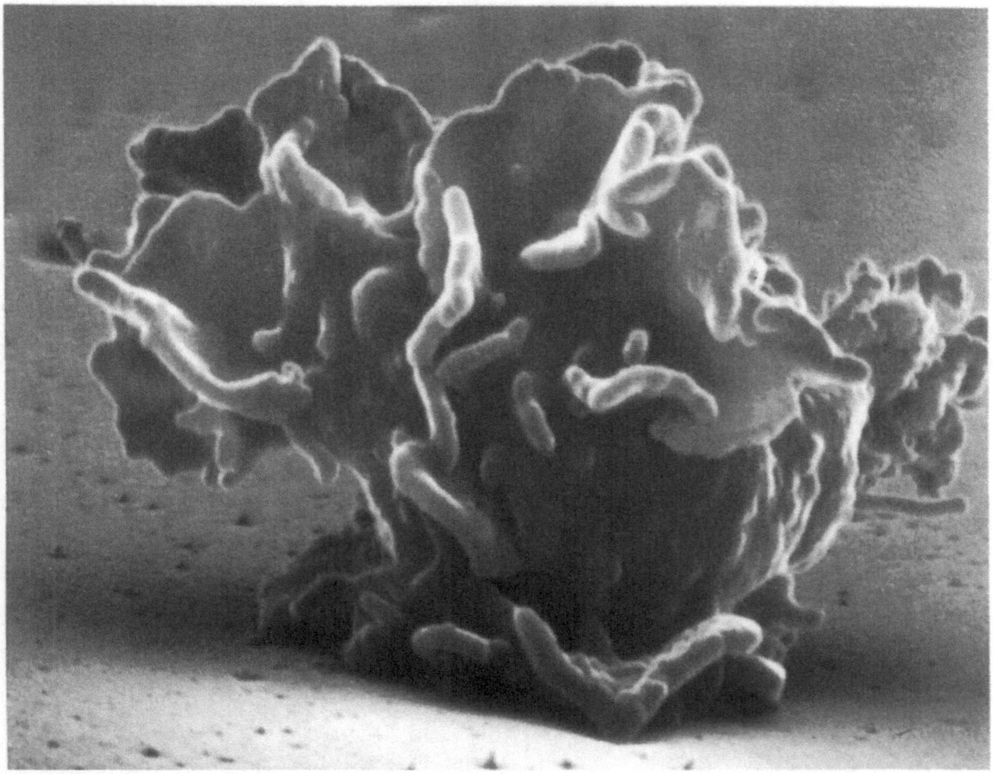
a

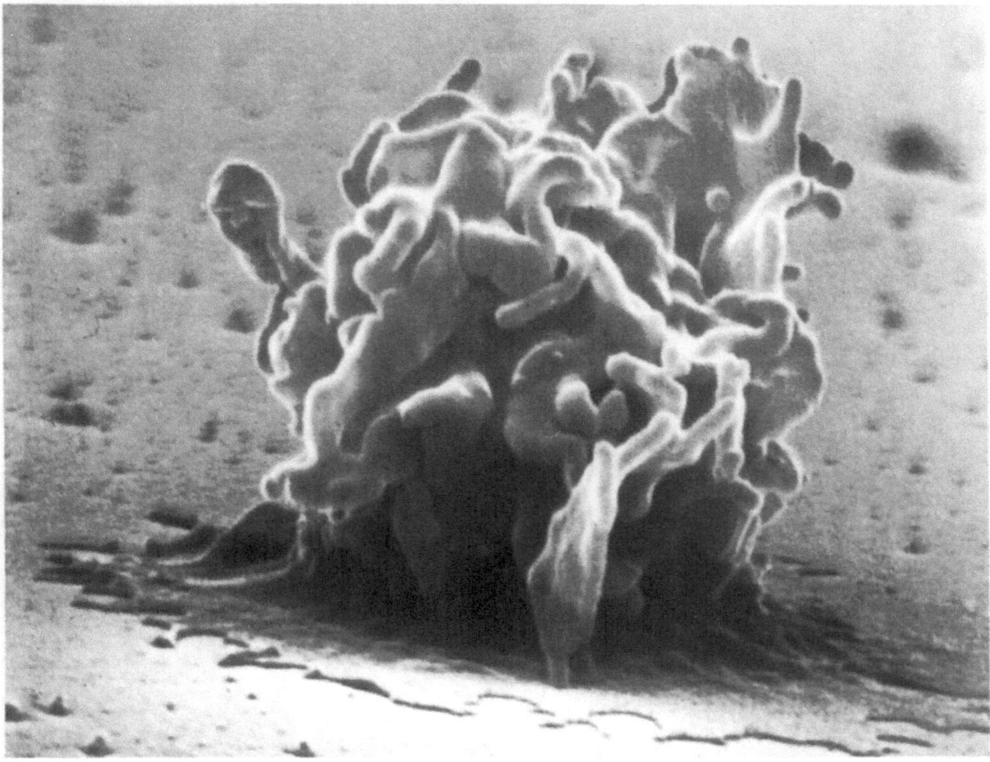
b

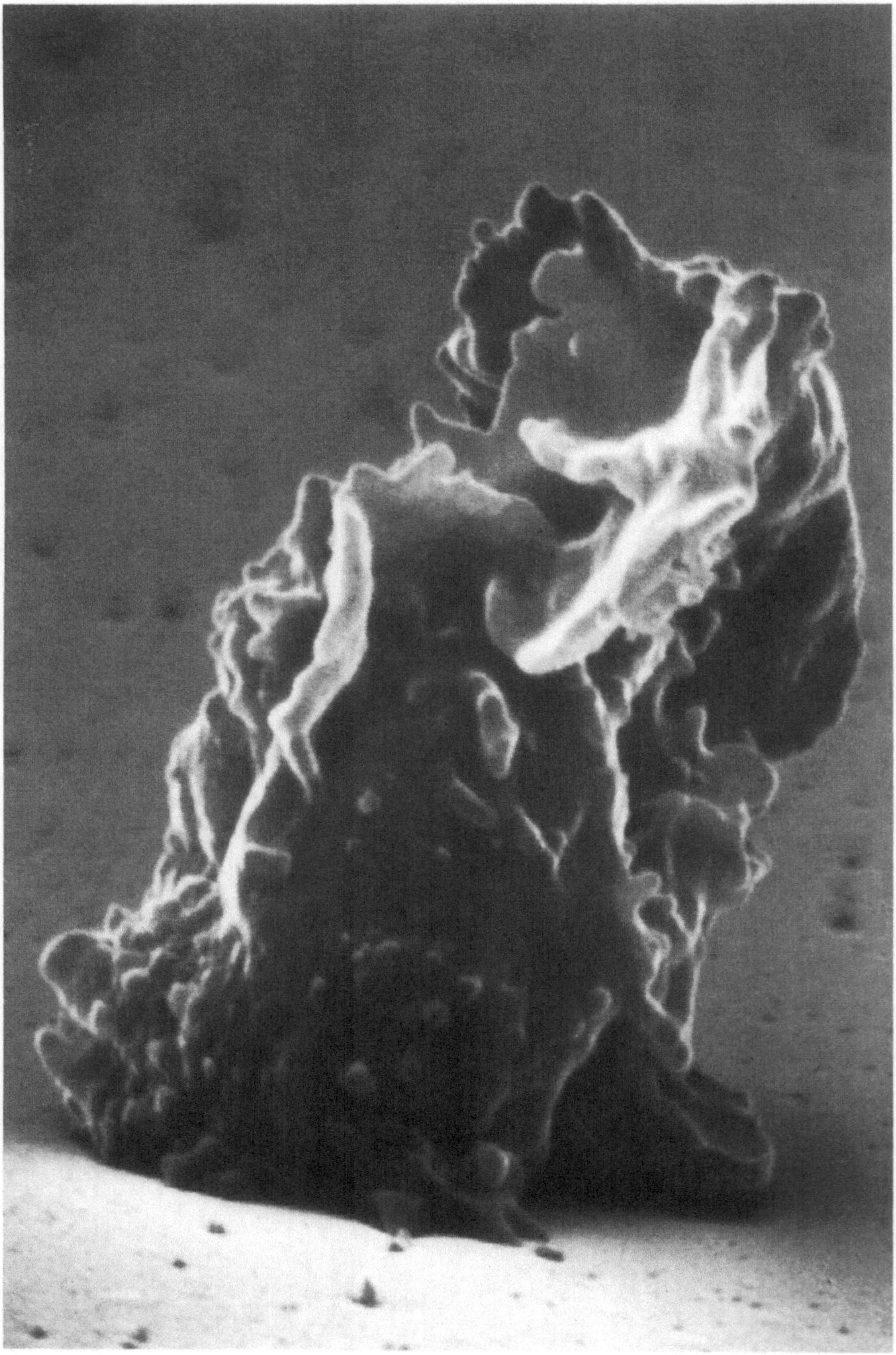

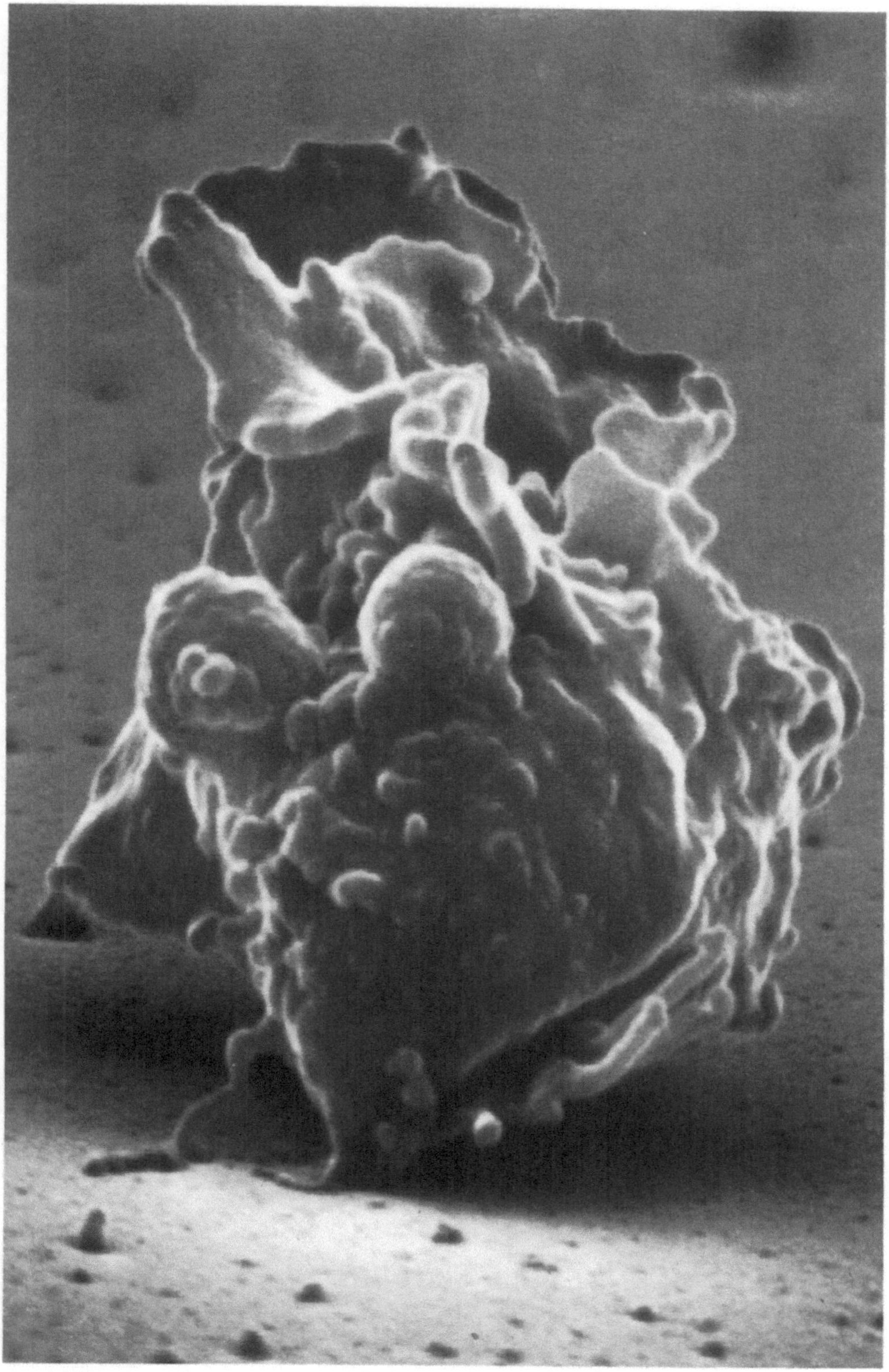

Figure 97. Leukemia cells adhering to other cells

Two leukemia cells, one spheric (a), the other polarized (b), are attached to cells with large veil-like extensions. The spheric cell is so firmly adherent that it is lifted off the substrate (a). The polarized cell is fastened by means of a foot-like extension (b)

a SEM: × 10,000
b SEM: × 10,500

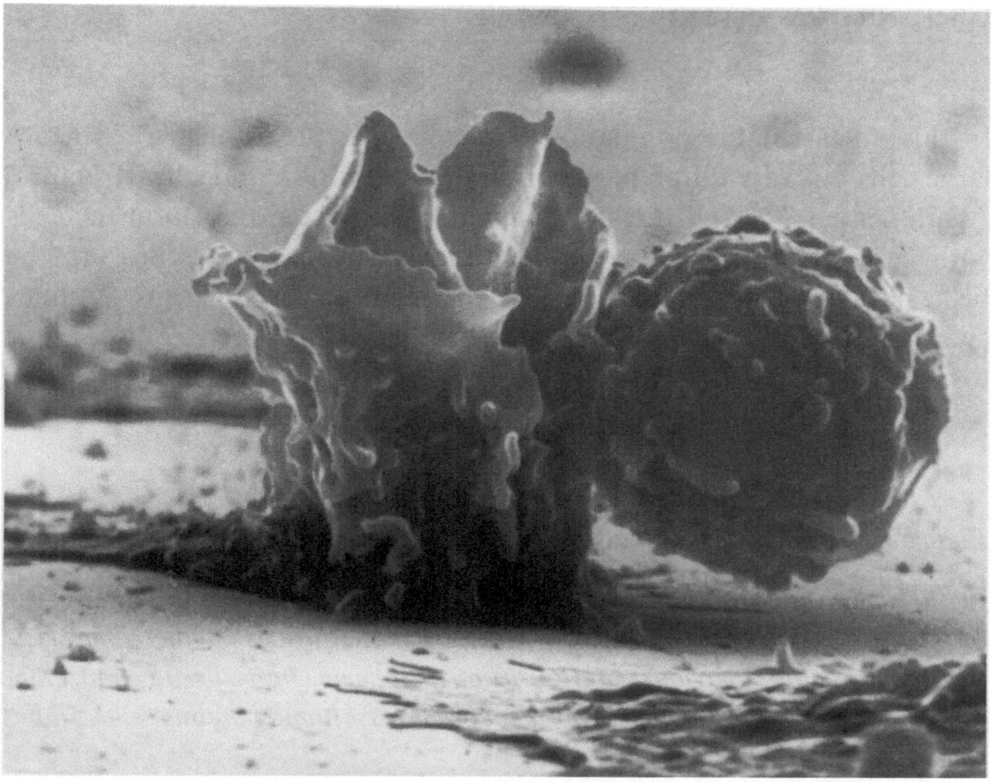

a

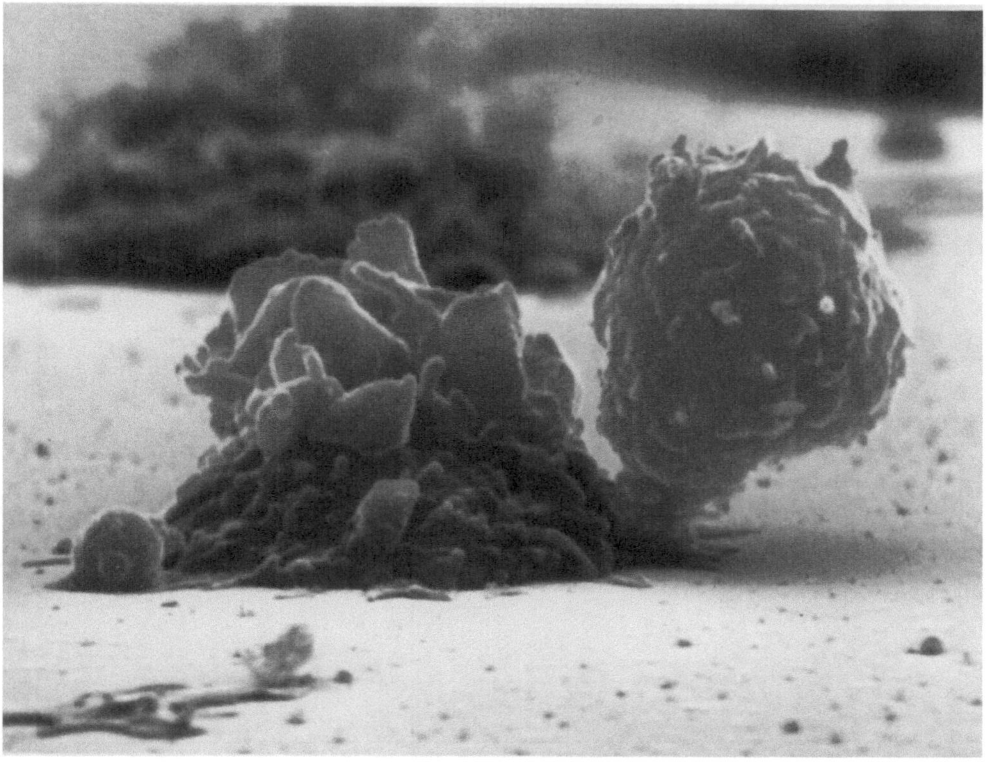

b

In this unclassifiable population two cell types predominate. Both are large cells but very different in surface configuration and ultrastructure. Again neither TEM, SEM, nor cinematography could position the cells with certainty. Both cell types are highly locomotive, following the blast pattern of movement. Besides the large cells, numerous granulocytes and lymphocytes are present.

Figure 98. Leukemic population

In these SE micrographs one of the two cell types is completely spheric (1) covered with short microvilli, while representatives of the other type (2) have assumed various configurations: spheric, half-spread, and elongated. The surface of the type 2 cell is less homogeneous and shows folds and ruffles

a SEM: ×3000
b SEM: ×2900

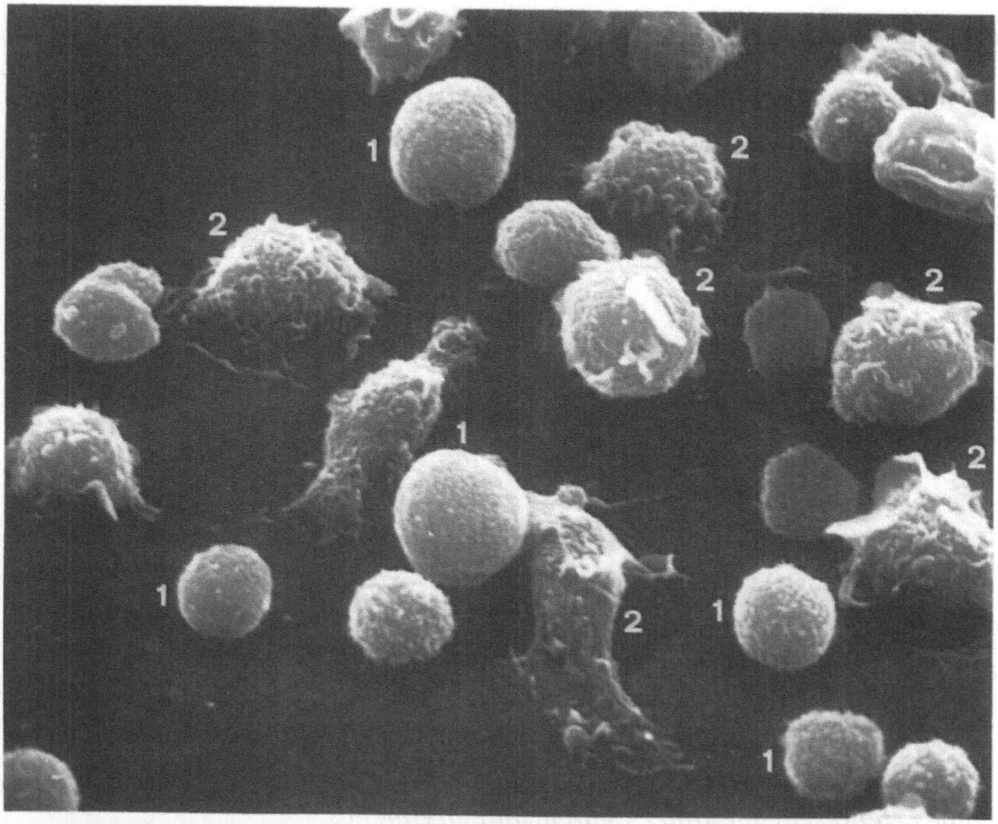

a

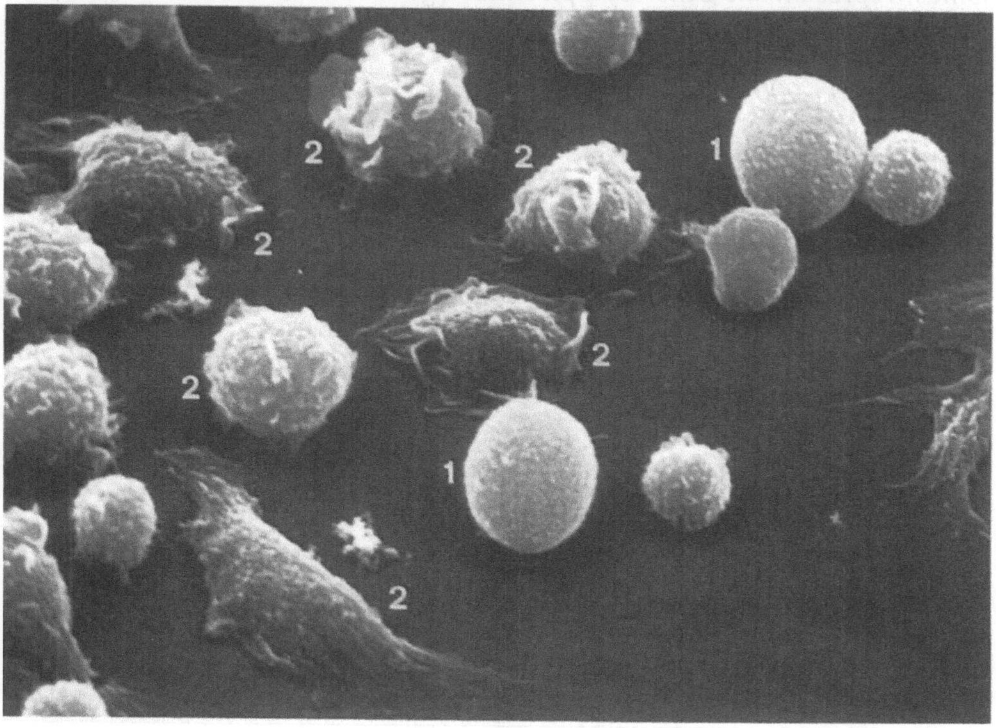

b

Figure 99. Leukemic cells: type 1

A higher magnification shows several cells from type 1 attached to either the glass surface or to a spread macrophage

SEM: ×5200

Figure 100. Leukemic cells: type 1 *(see page 166)*

The particular surface architecture of the type 1 cell, densely and evenly covered with stub-like microvilli, is shown in both the SE (a) and the TE micrograph (b). TEM (b) further reveals that the cytoplasm of the type 1 cell contains a large amount of rough endoplasmatic reticulum that can be arranged in the form of concentric cysternae (arrow). A possible resemblance to the endoplasmatic reticulum of immature plasma cells has been suggested by Anderson (1966) and Tanaka and Goodman (1972). A well-developed Golgi complex lies within a nuclear indentation. Round and oval mitochondria, together with free ribosomes and polyribosomes, represent the other cytoplasmic organelles

a SEM: ×14,200
b TEM: ×11,900

Figure 101. Leukemic cells: type 1 *(see page 167)*

Higher magnifications of SEM (a) and TEM (b) show details of the characteristic surface structure with its stub-like microvilli

a SEM: ×30,200
b TEM: ×27,800

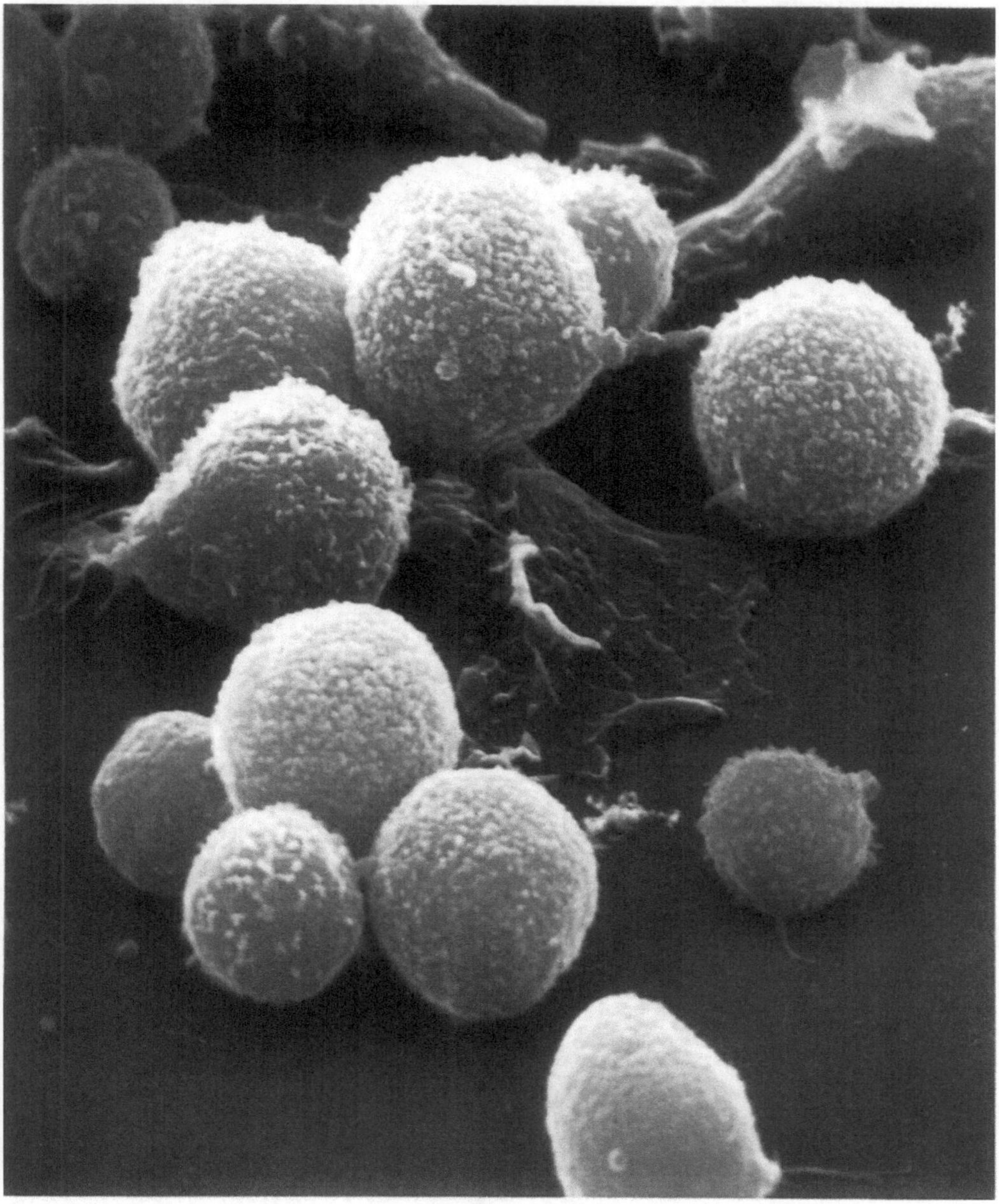

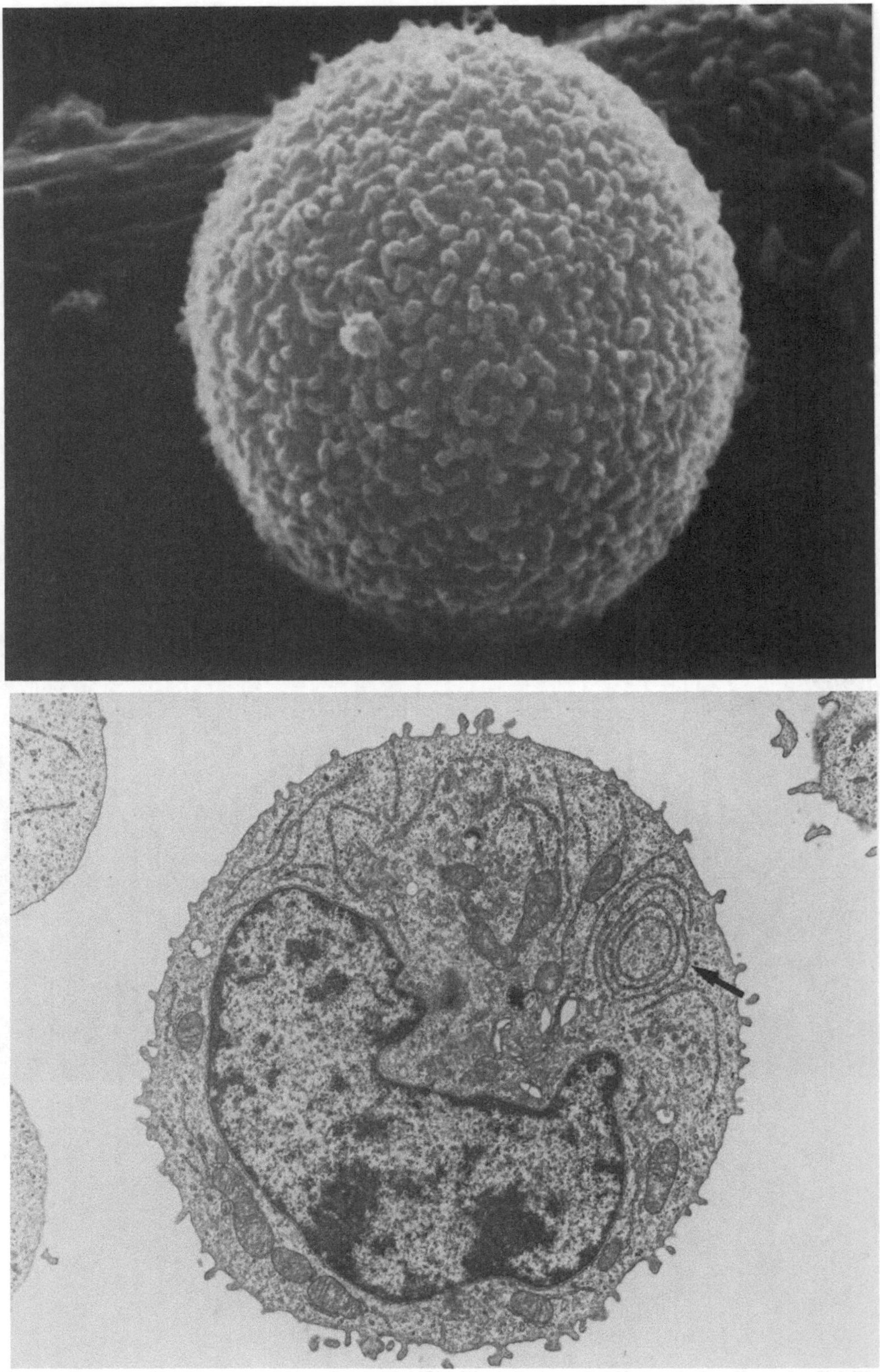

a

b

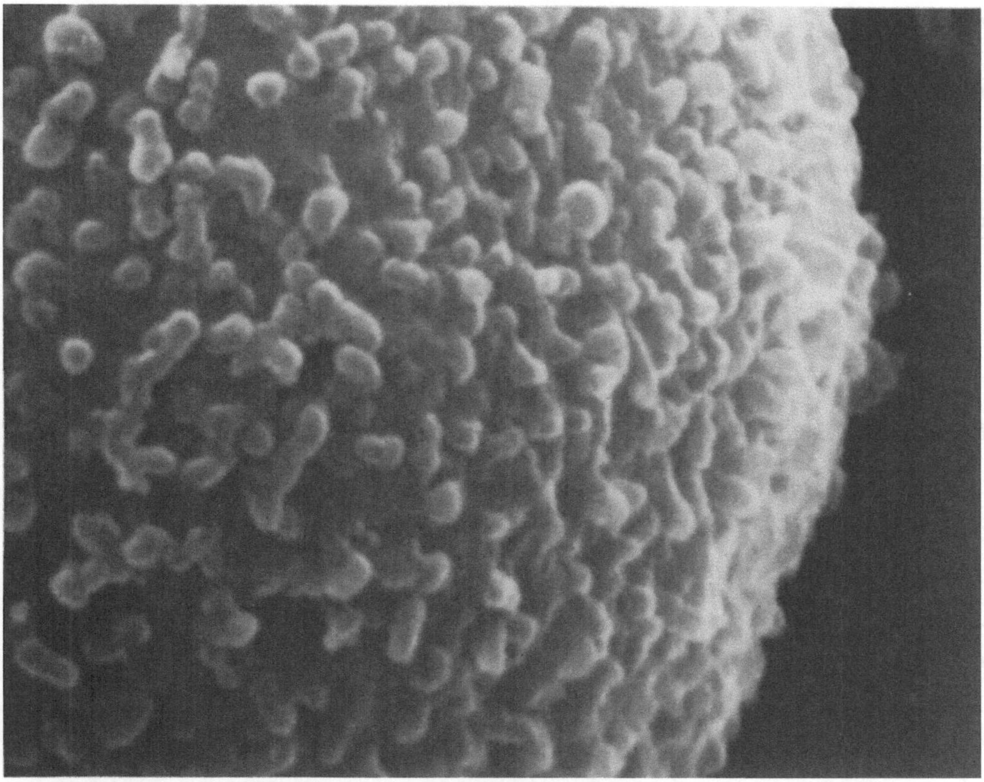
a

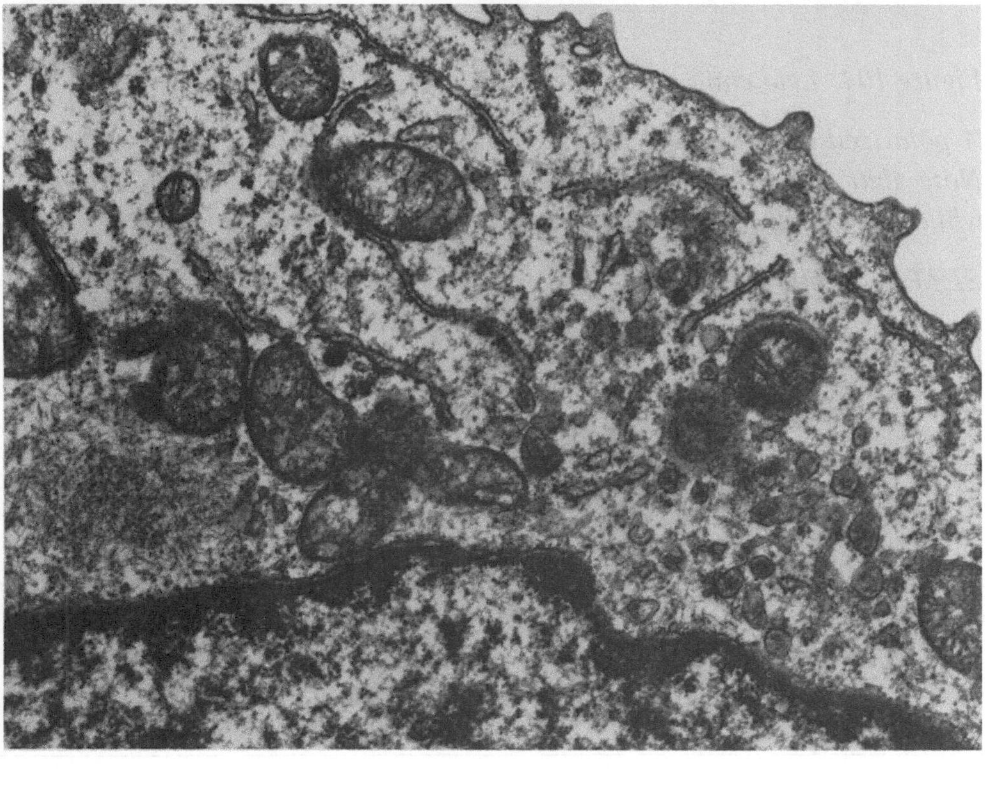
b

Figure 102. Leukemic cells: type 1

This picture demonstrates that cell type 1 not only exists in a spheric form but also that it can spread. Note that the typical surface features of the round cell are retained after spreading

SEM: × 8800

Figure 103. Leukemic cells: type 1 *(see page 170)*

A type 1 cell in a polarized configuration with large veils at the anterior part indicative of locomotion (a). Time-lapse film observations revealed many of these cells during locomotive activity (b)

a SEM: × 10,300
b Phase contrast: × 2600

Figure 104. Leukemic cell: type 1 *(see page 171)*

A polarized cell of type 1 adheres with its anterior part to the substrate. Note that the front is covered with the characteristic short microvilli, while those of the opposite part are longer

SEM: × 20,600

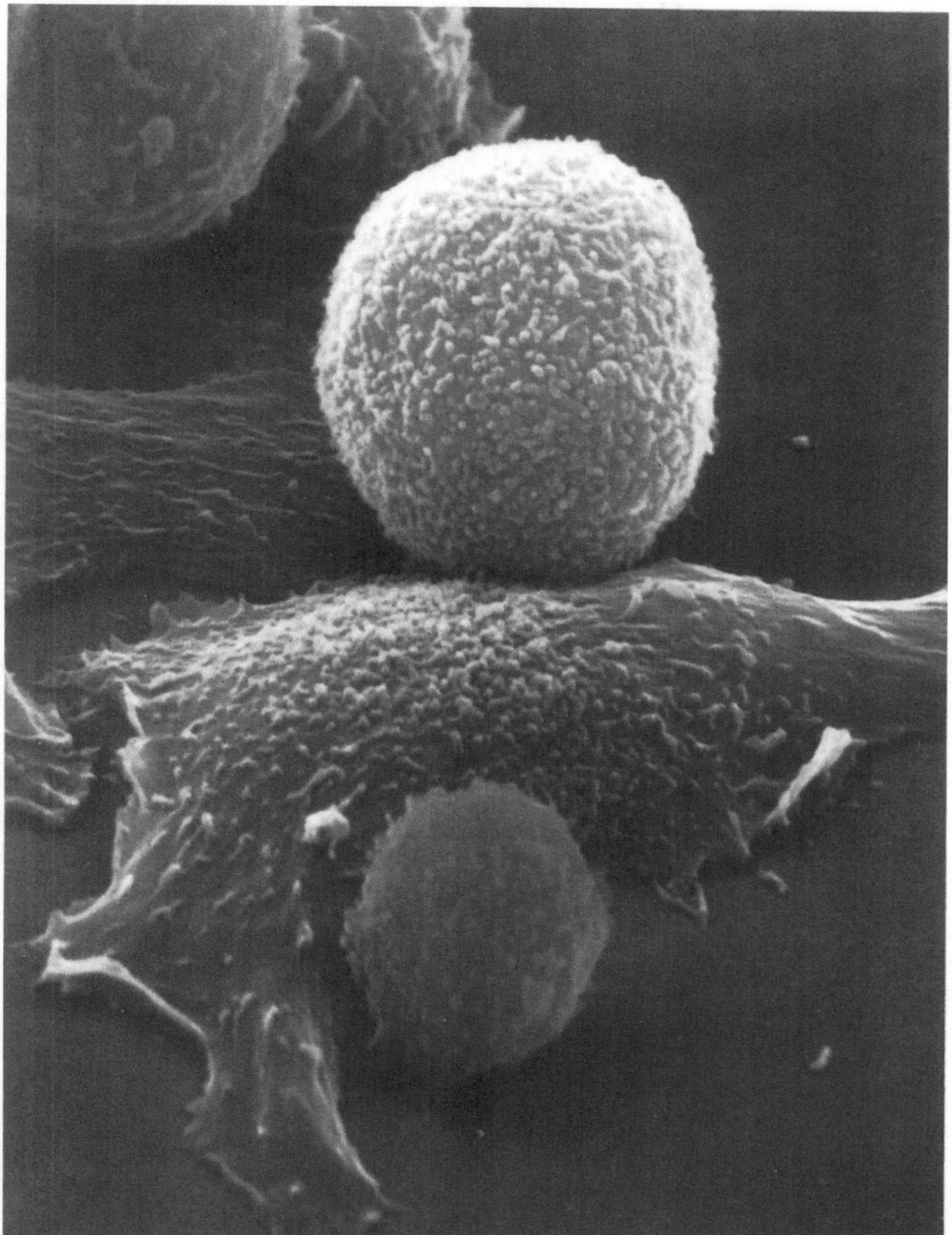

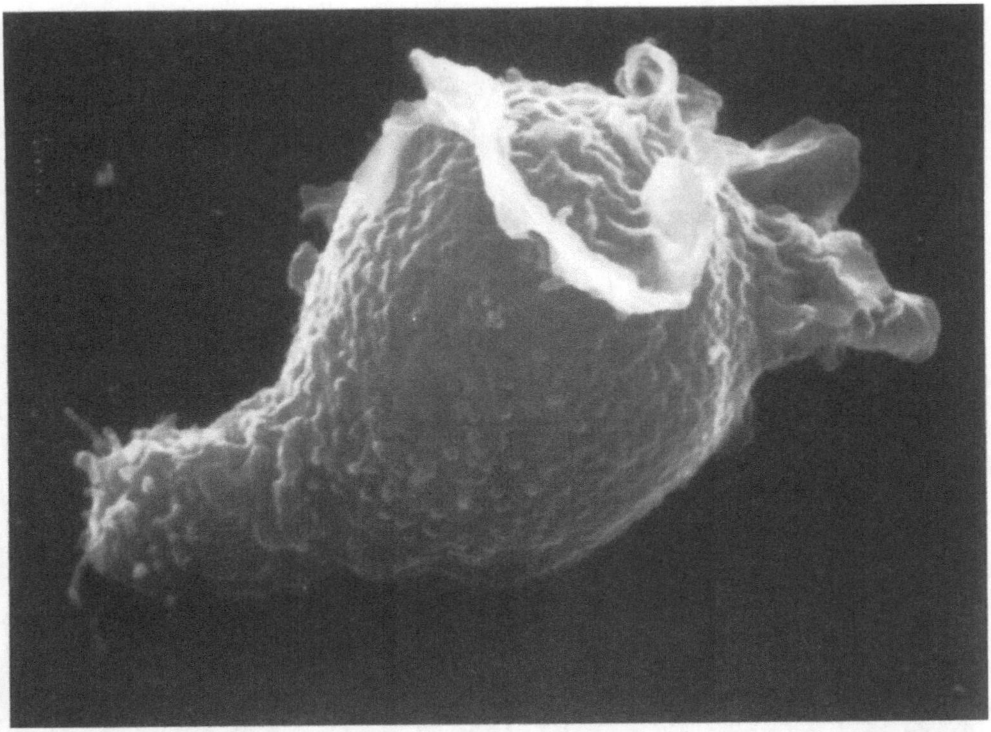

a

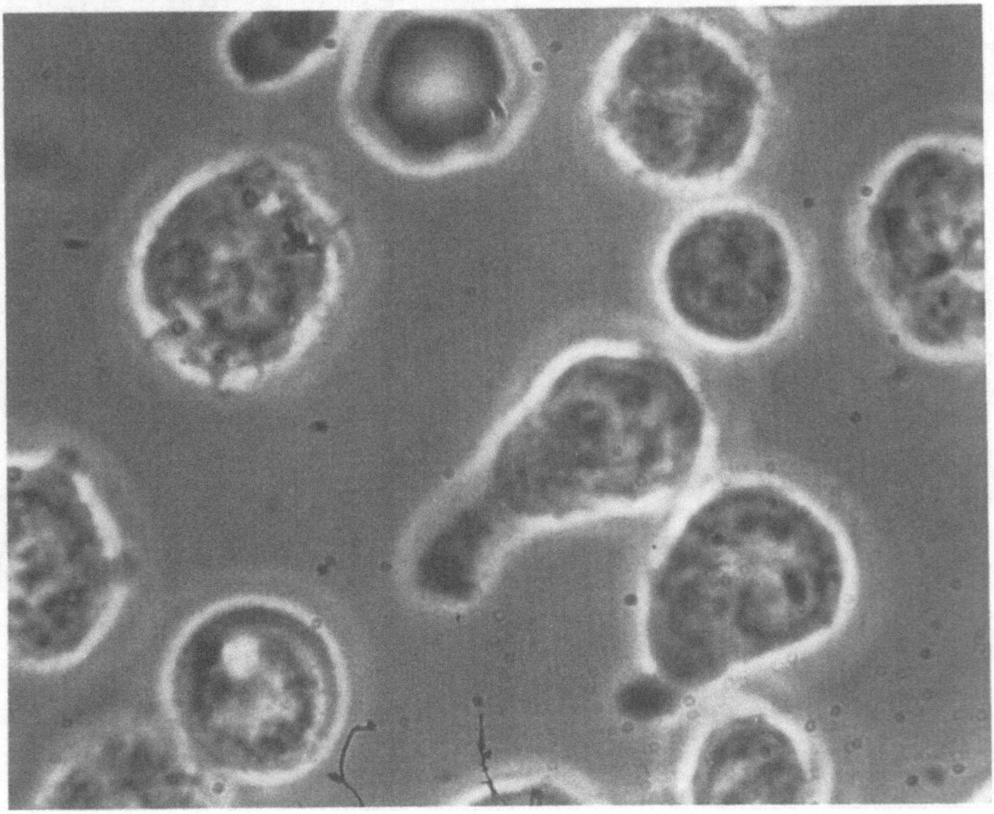

b

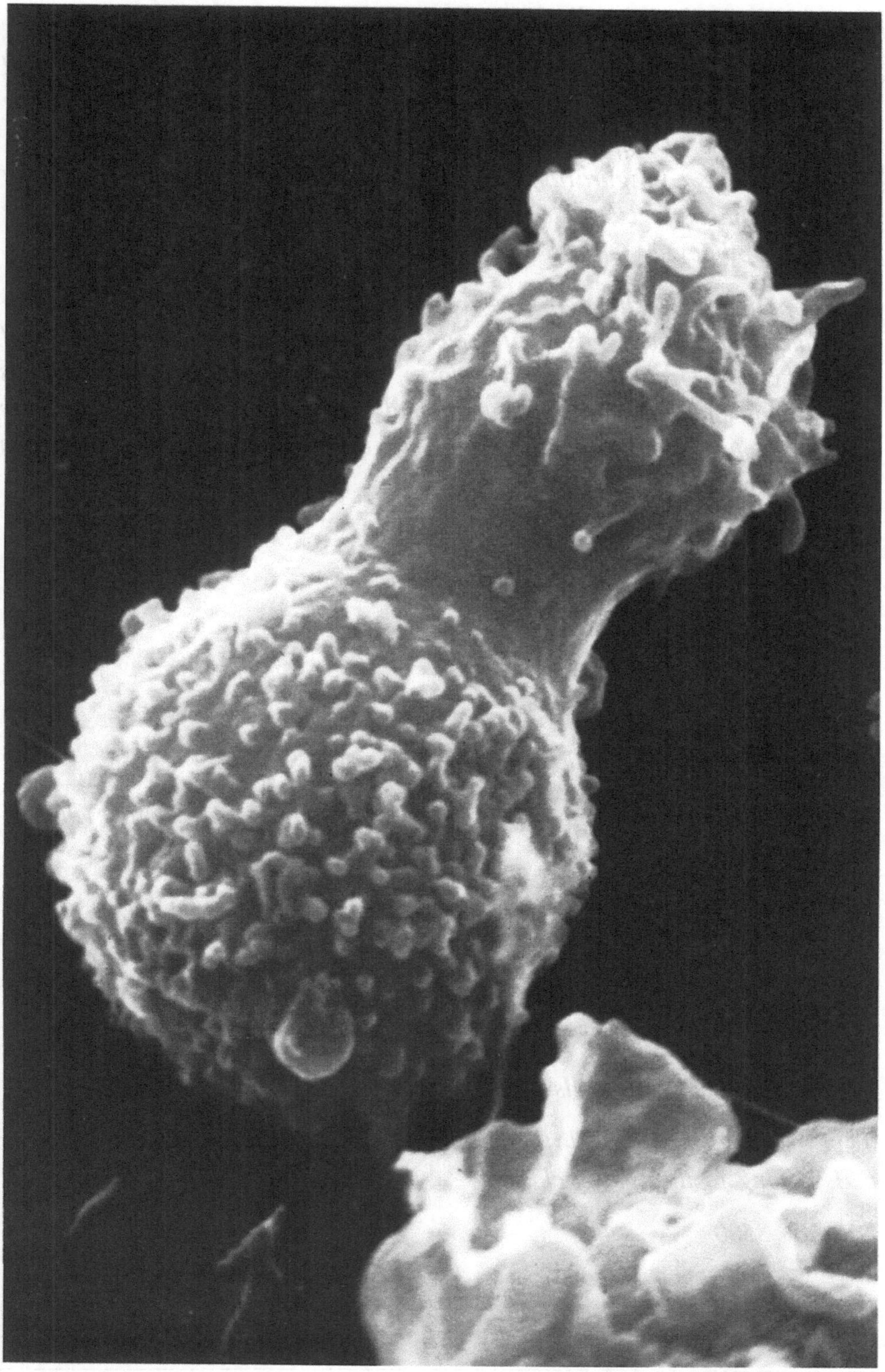

Figure 105. Leukemic cells: type 2

a *The completely different surface structure of the type 2 cell is well documented by SEM. Instead of the stub-like microvilli of the type 1 cells, those of type 2 have prominent folds and veil-like extensions. This surface configuration suggests a monocytoid nature. The large size of the cells is made especially obvious by the presence of the two thrombocytes (th)*

b *The monocytoid nature of type 2 cells is also recognizable in TE micrographs: The nucleus is contorted and, besides a variable number of small, dense granules, the cytoplasm contains short, irregular strands of rough endoplasmatic reticulum and free ribosomes. The similarity of the long cytoplasmic extensions to those in the SEM photograph (a) is evident*

a SEM: $\times$ *8000*
b TEM: $\times$ *6800*

Figure 106. Leukemic cell: type 2 *(see page 174)*

Type 2 cells can also be polarized while maintaining their surface characteristics

SEM: $\times$ *11,300*

Figure 107. Leukemic cells: type 2 *(see page 175)*

Type 2 cells can assume different shapes while spreading. The two spheric cells belong to the type 1 class

SEM: $\times$ *3500*

a

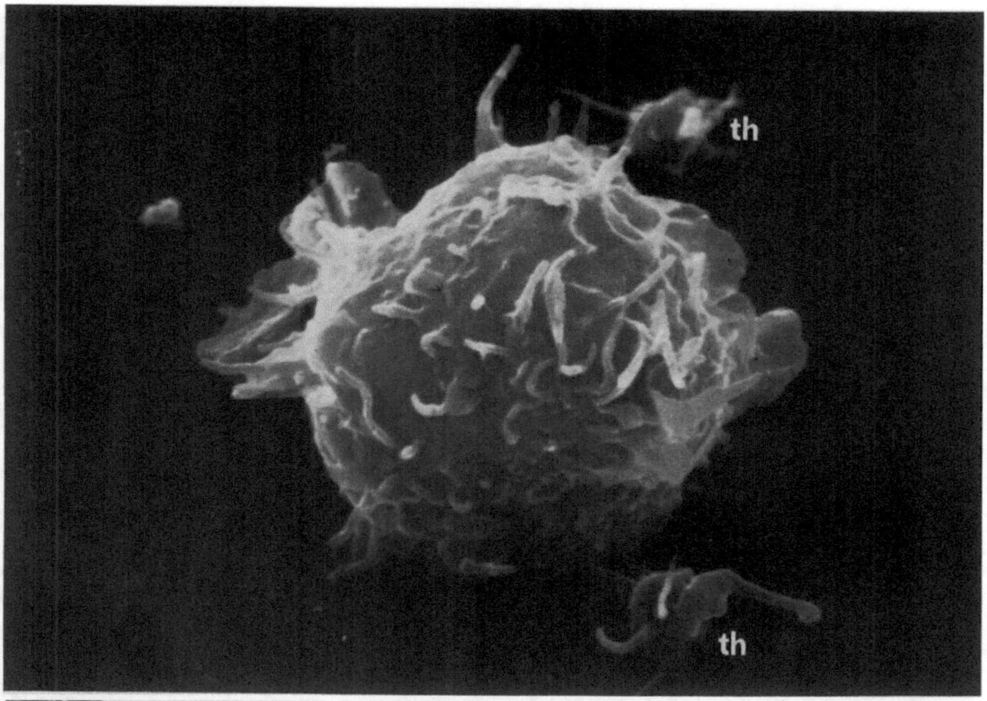

b

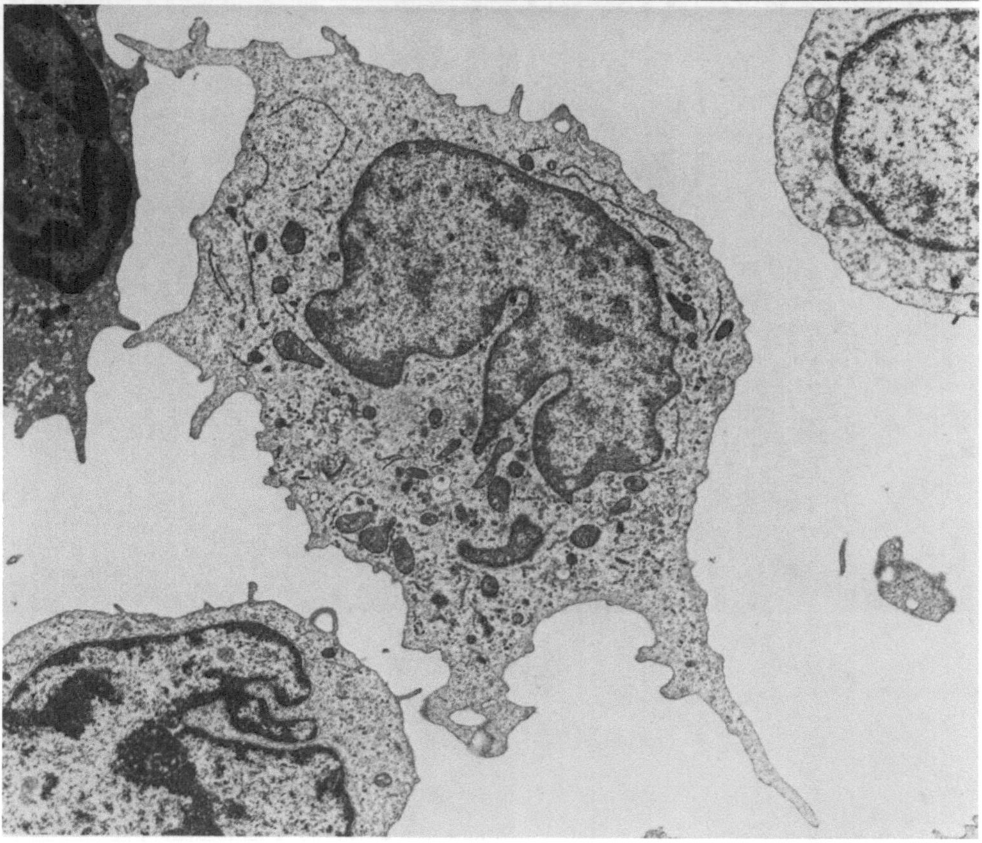

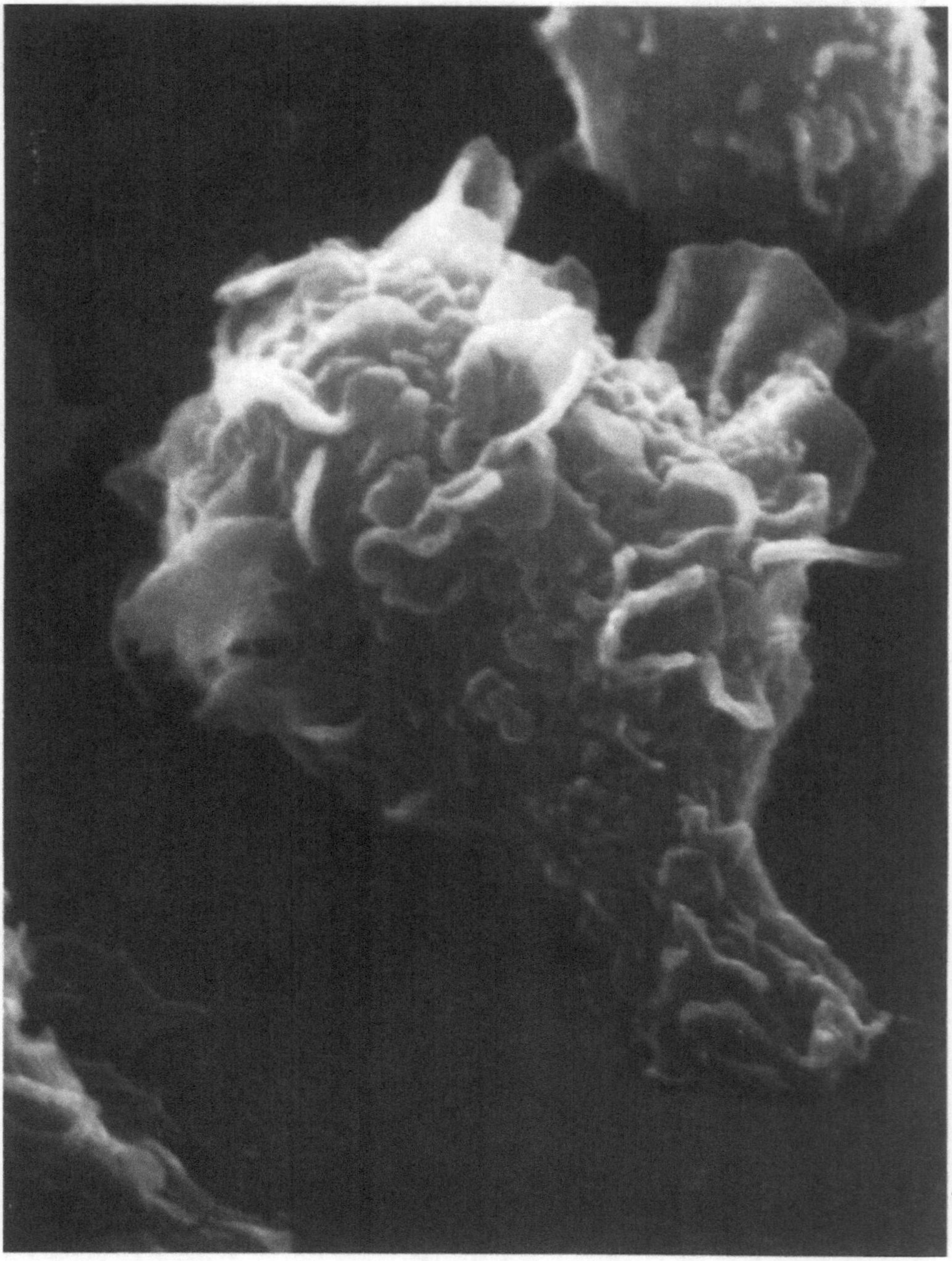

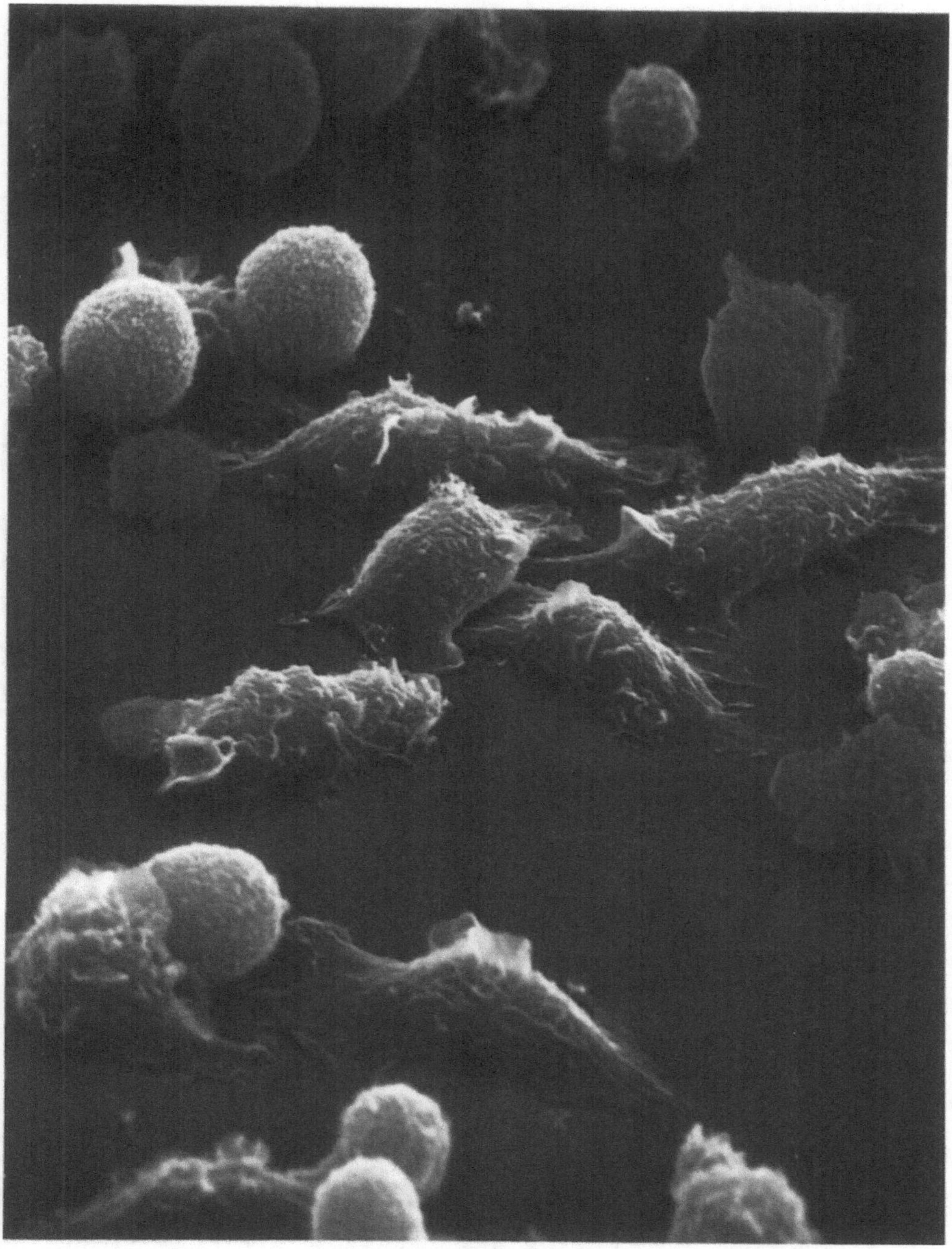

Summing-Up

Dynamic Morphology: What Can It Achieve?

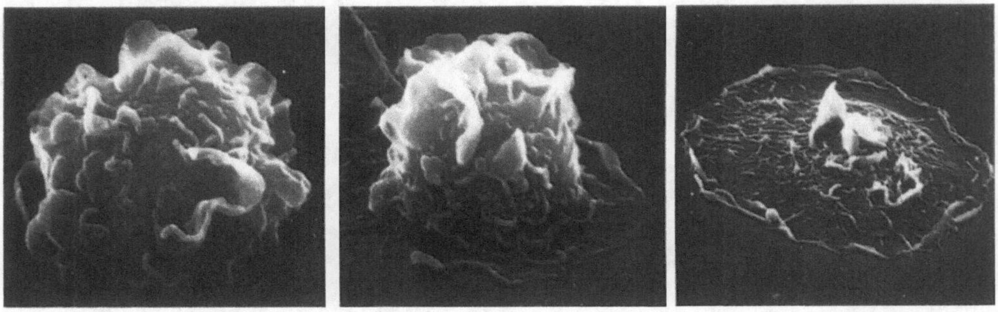

Dynamic morphology *can* recognize monocytes, both in the spherical and in the spread form.

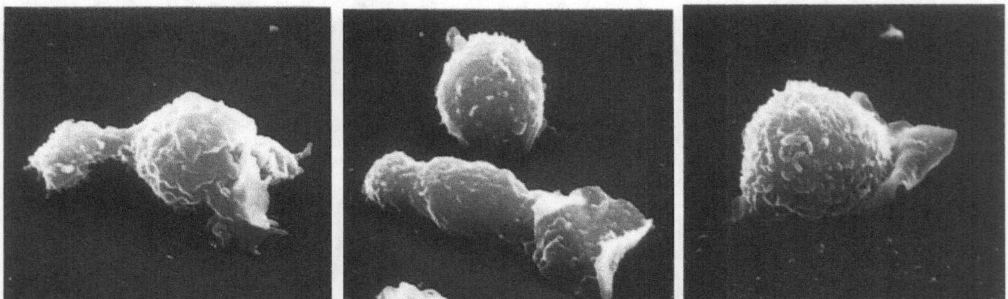

Dynamic morphology *can* recognize blast cells, promyelocytes, and myelocytes as soon as they have become locomotive. Locomotion is the key factor for the individually different configuration of the three cell types.

Dynamic Morphology: What Can It Not Achieve?

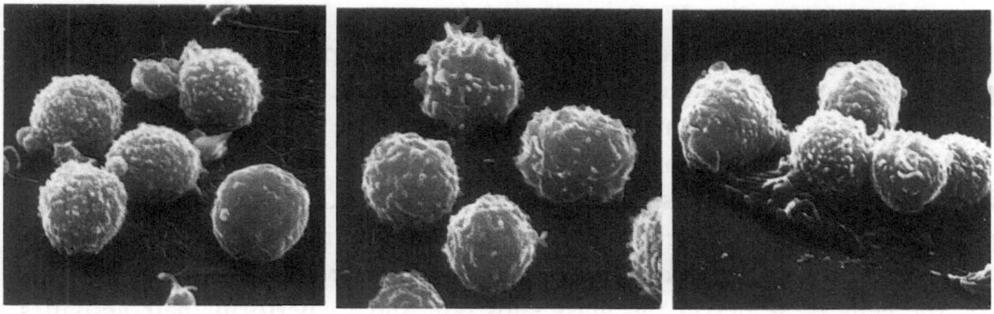

Dynamic morphology *cannot* determine whether cells in the spheric form are immature or mature, of lymphoid or of myeloid origin. Surface architecture of individual cells of one population can vary to such an extent that a classification for different populations is impossible.

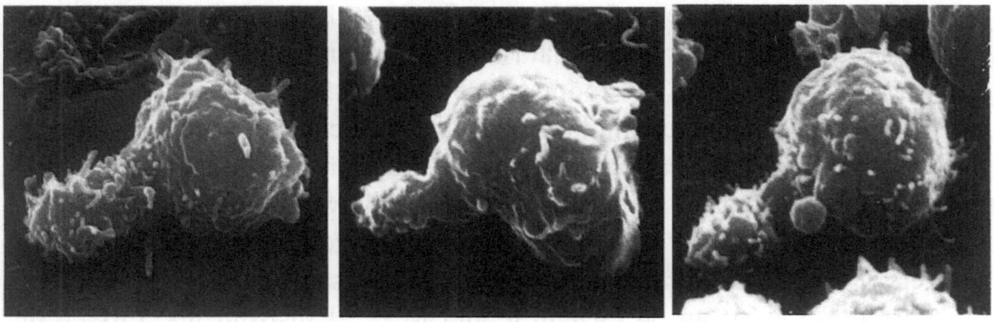

Dynamic morphology *cannot* distinguish whether locomotive blast cells are of myeloid or of lymphoid origin, or unclassifiable. Differences in surface architecture are not sufficient, and all blast cells move in the "blast pattern of locomotion."

Methodology

Rat leukemia cells were harvested from the peritoneal cavity, suspended in medium (BME, supplemented with 20% fetal calf serum, 0.03% glutamine and Hepes in a 0.01 *M* concentration), and immediately introduced into assembled culture chambers [Rose, 1954; Dvorak and Stotler, 1971]. The chambers were then transfered to 37° C.

From the heparinized venous blood from leukemia patients, buffy coats were prepared by a modification of the Ficoll gradient technique described by Böyum (1968). After centrifugation the cells were suspended in the same medium as used for the animal cells, washed three times, and then injected into the culture chambers which were kept at 37° C. While microcinematography started immediately and continued for hours, the cells intended for SEM remained in the chambers for $1^1/_2$ h. This period provided ample time for the cells to settle on the glass substrate and to perform their activities.

Preparation for SEM

After $1^1/_2$ h, the chambers were disassembled, the bottom glass slides rinsed in 0.1 *M* phosphate buffer (pH 7.3), and fixed for 1 h in 2% glutaraldehyde. This was followed by four additional rinses in the same buffer and by postfixation for 1 h in 1% OsO_4 at room temperature. After rapid dehydration in graded series of ethanol, ethanol/amyl acetate, and concentrated amyl acetate, the specimens were critical point-dried in a Polaron E 3000 (Polaron Equipment Ltd., London, UK) using carbon dioxyde following the method described by Anderson (1951). The cells were covered with gold using a sputter coater (Balzers Union AG, Balzers, Liechtenstein). The samples were examined in a Cambridge Stereoscan S4 (Scientific and Medical Instruments, Ltd., Cambridge, UK), operated at an accelerating voltage of 20 or 30 kV with a tilt angle of 60° and 85°.

Preparation of Bone Marrow for SEM

For the preparation of the bone marrow from BD IX rats, injected intravenously with L 5222 cells, a perfusion fixation technique was utilized. Anesthetized animals were perfused 1–2 min via the thoracic aorta with a Ringer solution containing 0.1% procaine and 0.2% Macrodex. This was followed by a perfusion for 12 min with a mixture of 2% gluteraldehyde and 2% paraformaldehyde in 0.1 *M* phosphate buffer (pH 7.3). The perfusion flow amounted to 20 ml/min. After the bone marrow was removed from the femur, following the technique given by de Bruyn et al. (1971) to avoid mechanical disruption of the tissue, it was further fixed for 2–4 h in the same solution used for perfusion. Afterwards the tissue was rinsed in buffer, postfixed in 2% osmium tetroxyde, and dehydrated in a graded series of ethanol. It was then immersed in liquid nitrogen and cryofractured as described by Humphreys et al. (1974). The resulting fragments were thawed in absolute ethanol before they were further dehydrated in amyl acetate, critical point-dried, and coated with gold as described for single cells. For these preparations the Cambridge Stereoscan S4 was used with an accelerating voltage of 10 or 20 kV and a specimen tilt of 30°.

Preparation for TEM

The leukemia cells were suspended in the same medium as described before and kept for 15 min at 37° C. Double fixation (2% glutaraldehyde, 1% osmium tetroxyde in 0.1 *M* cacodylate buffer) was carried out at room temperature. The cells were then block-stained in a 0.5% alcoholic solution of uranyl acetate, dehydrated in a graded series of ethanol, passed through propylen oxide, and embedded in Epon. Thin sections were examined in a Siemens 102 electron microscope.

Microcinematography

The cells were observed and filmed in culture chambers using an inverted Wild phase-contrast microscope and an inverted Leitz microscope (Diavert) equipped with Nomarski contrast optics and housed in temperature-controlled boxes of 37° C. Each microscope was connected to a Bolex H 16 camera and a Paillard Wild Variotimer.

The effect of light energy was kept minimal by the presence of an electromagnetic shutter placed between light source and condensor. Kodak Plus X reversal 16-mm black and white film was exposed in intervals of 2–4 s (15–30 frames/min), depending on the activity of the individual cell populations.
For analysis of speed, duration of locomotion, and length of the pathway, the films were projected on a paper of 21 × 28 cm and the outlines of locomotive cells drawn by hand at always the same time interval. Changes of shape, occurring during on-spot motility or during locomotion, were recorded by drawings of the same cell at intervals of 15 s.

Cells intended for filming on a natural substrate were injected into culture chambers in which part of isologous mesentery, mounted on a glass slide, had been placed.

Hematologic Data

The hematologic data were kindly provided by the Department of Internal Medicine and the Children's Hospital, both from the University of Zürich.

Acute Myeloid Leukemias (AML)

Case 1.[a] Female, age 48

	Before treatment (%)	40 h after beginning of therapy (%)
Myeloblasts	88	57
Promyelocytes	9	1
Neutrophils	1	16
Monocytes	0	4
Lymphocytes	2	22

Case 2.[a] Male, age 74

	Before treatment (%)
Myeloblasts	93
Myelocytes	1.5
Neutrophils	1
Monocytes	1
Lymphocytes	3.5

Case 3.[b] Male, age 25

	Before treatment (%)
Paraleukoblasts	2
Neutrophils	49
Basophils	0.5
Monocytes	1
Lymphocytes	47.5

Case 4.[b] Male, age 6

	Before treatment (%)
Myeloblasts	61.6
Promyelocytes	2.8
Neutrophils	5.8
Lymphocytes	29.8

Case 5.[a] Male, age 75

	Before treatment (%)
Myeloblasts	67.5
Promyelocytes	12.5
Myelocytes	3.5
Metamyelocytes	1
Neutrophils	5
Eosinophils	3.5
Lymphocytes	7

[a] Department of Internal Medicine
[b] Children's Hospital

Case 6.[b] Female, age 7

	Before treatment (%)
Myeloblasts	31
Promyelocytes	13
Myelocytes	11
Neutrophils	4
Eosinophils	24
Monocytes	11
Lymphocytes	6

Chronic Myeloid Leukemia (CML)

Case 1.[a] Female, age 77

	Before treatment (%)
Myeloblasts	8.5
Promyelocytes	11
Myelocytes	19.5
Metamyelocytes	4
Neutrophils	36.5
Eosinophils	1.5
Basophils	3
Monocytes	0.5
Lymphocytes	15.5

Acute Myelomonocytic Leukemias (AMML)

Case 1.[a] Male, age 31

	Before treatment (%)
Myeloblasts	68
Metamyelocytes	0.5
Neutrophils	9
Monocytes	10.5
Lymphocytes	12

Case 2.[a] Male, age 18

	Before treatment (%)
Myeloblasts	82
Promyelocytes	0.5
Myelocytes	2
Metamyelocytes	0.5
Neutrophils	5.5
Eosinophils	1
Lymphocytes	8.5

Case 3.[a] Male, age 72

	Before treatment (%)	5 months after beginning of therapy (%)
Myeloblasts	51	1
Promyelocytes	2	2
Myelocytes	4	22
Metamyelocytes	2	–
Neutrophils	12	21
Eosinophils	2.5	0.5
Basophils	1	0.5
Monocytes (atypical)	14	24.5
Lymphocytes	11.5	28.5
Plasma cells	0	0.5

[a] Department of Internal Medicine
[b] Children's Hospital

Case 4.[a] Male, age 43

	Before treatment (%)
Myeloblasts	46
Promyelocytes	1
Myelocytes	1
Neutrophils	2
Monocytes (atypical)	32
Lymphocytes	18

Case 5.[a] Male, age 42

	Before treatment (%)
Monocytoid blasts (atypical)	76
Metamyelocytes	1.5
Neutrophils	6
Monocytes	5.5
Lymphocytes	11

Acute Lymphoid Leukemias (ALL)

Case 1.[b] Female, age 6

	Before treatment (%)
Paraleukoblasts	96.2
Promyelocytes	0.2
Neutrophils	0.2
Lymphocytes	3
Plasma cells	0.4

Case 2.[a] Male, age 59

	Before treatment (%)
Lymphoblasts	85
Neutrophils	4
Eosinophils	1
Monocytes	2
Lymphocytes	8

Case 3.[b] Male, age 3

	Before treatment (%)
Paraleukoblasts	96.8
Promyelocytes	0.2
Myelocytes	0.4
Neutrophils	1.2
Eosinophils	0.2
Monocytes	1.2

Case 4.[a] Male, age 26

	Before treatment (%)
Lymphoblasts	92
Neutrophils	0.5
Lymphocytes	7.5

Chronic Lymphoid Leukemia (CLL)

Case 1.[a] Female, age 41

	Before treatment (%)
Lymphoblasts	7
Neutrophils	1.5
Lymphocytes (partly atypical)	91.5

[a] Department of Internal Medicine
[b] Children's Hospital

Unclassifiable Leukemias

Case 1.[b] Male, age 3

	Before treatment (%)
Paraleukoblasts	23
Myeloblasts	0.3
Promyelocytes	1.7
Myelocytes	4.7
Neutrophils	35.7
Basophils	0.3
Monocytes	0.7
Lymphocytes	32.6
Plasma cells	1

Case 2.[a] Male, age 63

	Before treatment (%)
Blast cells (atypical)	75
Neutrophils	1
Monocytes	1
Lymphocytes (atypical)	22
Plasma cells	1

[a] Department of Internal Medicine
[b] Children's Hospital

Bibliography

Alexander, E., Sanders, S., Braylan, R.: Purported differences between human T- and B-cell surface morphology is an artefact. Nature **261**, 239–241 (1976)

Allen, T.D., Dexter, T.M.: Surface morphology and ultrastructure of murine granulocytes and monocytes in long-term liquid culture. Blood Cells **2**, 591–609 (1976)

Anderson, D.R.: Ultrastructure of normal and leukemic leukocytes in human peripheral blood. J. Ultrastruct. Res. Suppl. **9**, 5–42 (1966)

Anderson, T.F.: Techniques for the preservation of three-dimensional structure in preparing specimens for the electron microscope. Trans. N.Y. Acad. Sci. **13**, 130–134 (1951)

Bekkum, van D.W., van Oosterom, P., Dicke, K.A.: In vitro colony formation of transplantable rat leukemias in comparison with human acute leukemias. Cancer Res. **36**, 941–946 (1976)

Belpomme, D., Dantchev, D., Karima, A.M., Lelarge, N., Joseph, R., Caillou, B., Lafleur, M., Mathé, G.: Search for correlations between immunological and morphological criteria used to classify lymphoid leukemias and non-Hodgkin's hematosarcomas. In: Recent Results in Cancer Research, Vol. 56: Lymphocytes, macrophages, and cancer. Mathé, G., Florentin, I., Simmler, M.-C. (eds.). Berlin-Heidelberg-New York: Springer 1976, p. 131

Bessis, M.: Living Blood Cells and Their Ultrastructure. Berlin-Heidelberg-New York: Springer 1973

Bessis, M.: Commentary: Can one diagnose a cell type safely from its surface morphology? Blood Cells **2**, 607–608 (1976)

Bessis, M., de Boisfleury, A.: Les mouvements des leucocytes étudiés au microscope électronique à balayage. Nouv. Rev. Fr. Hématol. **11**, 377–400 (1971)

Bessis, M., de Boisfleury, A.: A catalogue of white blood cell movements (normal and pathologic). Blood Cells **2**, 365–410 (1976)

Boll, I.: Morphologische, phasenkontrast-kinematographische Studien zum Verhalten von Knochenmarkzellen in vivo. Blut **32**, 115–130 (1976)

Boll, I., Nitzel, H.: Die Kinetik leukämischer Blasten (Myeloblasten, Monoblasten, Lymphoblasten) im Vergleich mit Mononucleosezellen. In:

Leukämie. Gross, R., van de Loo, J. (eds.). Berlin-Heidelberg-New York: Springer 1972, p 199

Böyum, A.: Isolation of mononuclear cells and granulocytes from human blood. Scand. J. Clin. Lab. Invest. **19**, 77–89 (1968)

Bruyn, de P.P.H.: Locomotion of blood cells in tissue culture. Anat. Rec. **89**, 43–63 (1944)

Bruyn, de P.P.H.: The amoeboid movement of the mammalian leukocyte in tissue culture. Anat. Rec. **95**, 177–191 (1946)

Bruyn, de P.P.H., Michelson, S., Thomas, T.B.: The migration of blood cells of the bone marrow through the sinusoidal wall. J. Morphol. **133**, 417–438 (1971)

Catovsky, D., Frisch, B., van Noorden, S.: B, T and "null" cell leukaemias. Electron cytochemistry and surface morphology. Blood Cells **1**, 115–124 (1975)

Cohnen, G., Fischer, K., Ludwig, H., Brittinger, G.: Surface morphology of normal chronic lymphocytic leukaemia lymphocytes. Europ. J. Cancer **12**, 199–202 (1976)

Coleman, M.S., Greenwood, M.F., Hutton, J.J., Bollum, F.J., Lampkin, B., Holland, P.: Serial observations on terminal deoxynucleotidyl transferase activity and lymphoblast surface markers in acute lymphoblastic leukemia. Cancer Res. **36**, 120–127 (1976)

Colly, L.P., Töns, A., Hagenbeek, A.: Experimental chemotherapy in the BN myelocytic leukaemia. Personal communication

Dantchev, D., Belpomme, D.: Critical study of the mononuclear leukocyte morphology based on scanning electron microscopy in normal subjects and in patients with lymphoid or monocytoid proliferative disorders. Comparison with the T, B or null cell membrane phenotypes. Biomedicine **27**, 202–222 (1977)

Deegan, M.J., Cossman, J., Chosney, B.T., Schnitzer, B.: Hairy cell leukemia – Immunologic and ultrastructural study. Cancer **38**, 1952–1961 (1976)

Dvorak, J.A., Stotler, W.F.: A controlled-environment culture system for high resolution light microscopy. Exp. Cell Res. **68**, 144–148 (1971)

Fiebig, H.H., Zeller, W.J., Schmähl, D.: Development of meningeal leukemia in rats by intracerebral or intrameningeal inoculation of the acute transplantable leukemia L 5222. Z. Krebsforsch. **86**, 287–292 (1976)

Golomb, H.M., Reese, C.: Surface ultrastructural and marker characteristics of leukemic cells. In: Scanning Electron Microscopy. Johari, O., Becker, R.P. (eds.). Chicago, Illinois: IIT Research Institute 1976, part 5, p. 41

Golomb, H.M., Braylan, R., Polliack, A.: "Hairy" cell leukaemia (leukaemic reticuloendotheliosis): A scanning electron microscopic study of eight cases. Br. J. Haematol. **29**, 455–461 (1975a)

Golomb, H.M., Braylan, R., Reese, C., Variakojis, D., Brynes, R.K., Yachnin, S.: The Sézary syndrome cell: Surface ultrastructural characteristics. Acta Haematol. **54**, 106–114 (1975b)

Haemmerli, G., Felix, H.: Motility of acute human leukemia cells: A study by time-lapse cinematography and scanning electron microscopy. Blood Cells **2**, 415–430 (1976)

Haemmerli, G., Felix, H.: Dynamics of Leukemia Cells. (Film, 1977)

Haemmerli, G., Felix, H., Sträuli, P.: Motility of L 5222 rat leukemia cells. Virchows Archiv B [Cell Pathol.] **20**, 143–154 (1976)

Haemmerli, G., Felix, H., Sträuli, P.: Motility of L 5222 rat leukemia cells in the flattened state. Evidence against emperipolesis. Virchows Archiv B [Cell Pathol.] **24**, 165–178 (1977)

Haemmerli, G., Felix, H., Sträuli, P.: Dynamic morphology of human lymphoid leukemias. In: Recent Results in Cancer Research (in press) (1977)

Hagenbeek, A.: Extracorporeal irradiation of the blood in a rat leukaemia model. Doctoral thesis. Erasmus University of Rotterdam 1977

Hagenbeek, A., van Bekkum, D.W. (eds.): Proceedings of a workshop on "Comparative evaluation of the L 5222 and BN ML rat leukaemia models and their relevance for human acute leukemia." Leukemia Res., 1, No. 2/3 (1977)

Harriss, E.B., Hoelzer, D.: DNA synthesis time in leukaemia cells as measured by the double labelling and the percentage labelled mitoses methods. Cell Tissue Kinet. **4**, 433–441 (1971)

Harriss, E.B., Hoelzer, D.: Studies of the anaemia in an acute rat leukaemia. Br. J. Haematol. **26**, 593–604 (1974)

Hermens, A.F., Hagenbeek, A., van Bekkum, D.W., Zaat, G.M., Martens, A.C.M., Hoogenveen-van Beugen, E.: Analyses of cell kinetic parameters for an experimental BN ML leukemia transplantable into the inbred BN/Bi-Rij rat strain. Personal communication

Hoelzer, D., Calvo, W., Meyer-Hamme, K.D., Harriss, E.B.: Cell distribution and proliferation pattern of a transferable acute leukemia in rats. J. Natl. Cancer Inst. **50**, 1545–1553 (1973)

Hoelzer, D., Harriss, E.B., Jäger, C.H., Haas, R.J., Fliedner, T.M.: Effect of the acute rat leukemia L 5222 on bone marrow stroma cells. Cancer Res. **34**, 1892–1898 (1974)

Humphreys, W.J., Spurlock, B.O., Johnson, J.S.: Critical point drying of ethanol-infiltrated cryofractured biological specimens for scanning electron microscopy. In: Scanning Electron Microscopy. Johari, O., Corvin, I. (eds.). Chicago, Illinois: IIT Research Institute 1974, part 1, p. 275

Ivankovic, S., Zeller, W.J.: Leukämie L 5222 des Rattenstamms BD IX. Eine durch Aethylnitrosoharnstoff induzierte monozytäre myeloische, transplantierbare Form für zytochemische und chemotherapeutische Studien. Blut **28**, 288–292 (1974)

Martens, A.C.M., Hagenbeek, A.: Pulse cytophotometry of the BN myelocytic leukemia during development and during treatment with cytostatic drugs. Personal communication

Michaelis, T.W., Larrimer, N.R., Metz, E.N., Balcerzak, S.P.: Surface morphology of human leukocytes. Blood **37**, 23–30 (1971)

Müller-Hermelink, U., Müller-Hermelink, H.K.: Scanning electron microscopic investigations of acute leukemia. Virchows Archiv B [Cell Pathol.] **23**, 227–236 (1977)

Norberg, B., Rydgren, L., Söderström, N.: Lymphocyte locomotion. Morphological criteria of the direction of lymphocyte movement. Lymphology **6**, 187–191 (1973)

Norberg, B., Rydgren, L., Stenstam, M.: Amoeboid movement configuration. A cell configuration observed in tumour cells from 3 cases of bone marrow neoplasia. Scand. J. Haematol. **13**, 294–304 (1974)

Polliack, A.: Surface features of circulating human leukemic cells: Experience with 175 cases of leukemia examined by scanning electron microscopy. In: Scanning Electron Microscopy. Johari, O., Becker, R.P. (eds.). Chicago, Illinois: IIT Res. Inst. 1976, part 5, p. 31

Polliack, A.: Normal, Transformed and Leukemic Leukocytes. A Scanning Electron Microscopy Atlas. Berlin-Heidelberg-New York 1977

Polliack, A., de Harven, E.: Surface features of normal and leukemic lymphocytes as seen by scanning electron microscopy. Clin. Immunol. Immunopathol. **3**, 412–430 (1975)

Polliack, A., McKenzie, S., Gee, T., Lampen, N., de Harven, E.: Clarkson, B.D.: A scanning electron microscopic study of 34 cases of acute granulocytic, myelomonocytic, monoblastic histiocytic leukemia. Am. J. Med. **59**, 308–315 (1975a)

Polliack, A., Siegal, F.P., Clarkson, B.D., Fu, S.M., Winchester, R.J., Lampen, N., Siegal, M., de Harven, E.: A scanning electron microscopy and immunological study of 84 cases of lymphocytic leukaemia and related lymphoproliferative disorders. Scand. J. Haematol. **15**, 359–376 (1975b)

Polliack, A., Froimovici, M., Pozzoli, E., Lambertenghi-Deliliers, G.: Acute lymphoblastic leukemia. A study of 25 cases by scanning electron microscopy. Blut **33**, 359–366 (1976)

Pourreau-Schneider, N., Felix, H., Haemmerli, G., Sträuli, P.: The role of cellular locomotion in leukemic infiltration. An organ culture study on penetration of L 5222 leukemia cells into the chick embryo mesonephros. Virchows Archiv B [Cell Pathol.] **23**, 257–264 (1977)

Rasche, H., Hoelzer, D., Dietrich, M., Keller, A.: Hemostatic defects in experimental leukemia. Haemostasis **3**, 46–54 (1974)

Rich, A.R., Wintrobe, M.M., Lewis, M.R.: The differentiation of myeloblasts from lymphoblasts by their manner of locomotion. Bull. Johns Hopk. Hosp. **65**, 291–308 (1939)

Rose, G.: A separate and multipurpose tissue culture chamber. Tex. Rep. Biol. Med. **12**, 1074–1083 (1954)

Senda, N., Ueda, Y., Yoshitake, J., Takabayashi, H., Tamura, H., Inui, Y., Kishigami, Y.: The differentiation of leukemic cells on the basis of their movement. In: Annual Report of the Center for Adult Diseases. Imamura, A. (ed.). Osaka: Hamaya & Co. 1961, Vol. I, Part 3, p. 191

Tanaka, Y., Goodman, J. R.: Electron Microscopy of Human Blood Cells. New York: Harper & Row 1972

Tchernia, G., Grange, M.J., Gombaud-Saintonge, G.: Locomotion, necrotaxis and phagocytosis of normal and leukemic myeloblasts. Blood Cells **2**, 431–436 (1976)

Zeller, W.J., Ivankovic, S., Schmähl, D.: Chemotherapy of the transplantable acute leukemia L 5222 in rats. Cancer Res. **35**, 1168–1174 (1975)

Acknowledgements

We like to thank Susanne Baumgartner for her skillfull technical assistance, Beatrice Arnold and Charles Häberlin for their excellent photographic work, and Renée Baumann for the preparation of the manuscript.

For technical help at the Cambridge Stereoscan we thank Doris Metzger and Jean-Jacques Pittet from the Institute of Plant Biology, University of Zürich. Use of the Siemens electron microscope of the Institute of Brain Research, University of Zürich, was generously allowed by Prof. K. Akert.

We also like to thank the staff of the Department of Internal Medicine and of the Children's Hospital, both from the University of Zürich, for their friendly cooperation in supplying the leukemic blood samples.

The important financial support of our experimental work by the Zürich Cancer League and by the Swiss Cancer League is gratefully acknowledged.

Lymphocytes, Macrophages, and Cancer

Editors: G. Mathé, I. Florentin, M.-C. Simmler

53 figures. IX, 160 pages. 1976
(Recent Results in Cancer Research, Volume 56)
ISBN 3-540-07902-5

D. Metcalf

Hemopoietic Colonies

In Vitro Cloning of Normal and Leukemic Cells

54 figures, 28 tables. IX, 227 pages. 1977
(Recent Results in Cancer Research, Volume 61)
ISBN 3-540-08232-8

K. Lennert

Malignant Lymphomas

Other than Hodgkin's Disease

Histology-Cytology-Ultrastructure-Immunology

In collaboration with N. Mohri, H. Stein, E. Kaiserling, H.-K. Müller-Hermelink

220 figures, some in color. Approx. 800 pages. 1977

(Handbuch der speziellen pathologischen Anatomie und Histologie Band 1, Teil 3 B)
ISBN 3-540-08020-1

Red Cell Shape

Physiology, Pathology, Ultrastructure
Editors: M. Bessis, R.I. Weed, P.F. Leblond
Proceedings of a Symposium held at the Institute of Cell Pathology, Hôpital de Bicêtre, Paris, France June 20-21, 1972

147 figures. VIII, 180 pages. 1973
ISBN 3-540-06257-2
Distribution rights for Japan: Maruzen Co.Ltd., Tokyo

Unclassifiable Leukemias

Proceedings of a Symposium held October 11-13, 1974
Editors: M. Bessis, G. Brecher at the Institute of Cell Pathology, Hôpital de Bicêtre, Paris, France

81 figures, 1 color-plate, 38 tables.
VI, 270 pages. 1975
ISBN 3-540-07242-X

This monograph comprises the first issue of the international journal Blood Cells.

Springer-Verlag
Berlin
Heidelberg
New York

MIX
Papier aus verantwortungsvollen Quellen
Paper from responsible sources
FSC® C105338

If you have any concerns about our products,
you can contact us on
ProductSafety@springernature.com

In case Publisher is established outside the EU,
the EU authorized representative is:
Springer Nature Customer Service Center GmbH
Europaplatz 3, 69115 Heidelberg, Germany

Printed by Libri Plureos GmbH
in Hamburg, Germany